Finishing Touches:

An Insightful Look Into the Mirror of Aging

Praise for *Finishing Touches: An Insightful Look Into the Mirror of Aging*

"...a sensitive and thoughtful "journey inward" by a woman facing aging. Her balanced and personal view of the "glass half empty/half full" illuminates the myths and realities of aging. This book will be read and reread by elder women and their families with pleasure and appreciation."

~ Lucy Scott, Ph.D. Principal author, *Wise Choices Beyond Midlife: Women Mapping the Journey Ahead*

"*Finishing Touches* deals with the impact of aging on human relationships in a candid, and often touching manner. It reflects both the challenges and rewards of growing older."

~ Laura Impastato, Editor, *Senior World Magazine*, California

"*Finishing Touches* is so full of illumination that you'll feel you've been helped to understand the aging process better, and that you've grown in wisdom as well. You'll say 'Why, that's me! That's us!" at page after page."

~ Eugenie H. Wheeler, MSW, Columnist, *Ventura County Star*

"Unlike technical books of "how-to," *Finishing Touches* is a book of "how-is," a calm reinvention of one woman's aging process. In this highly readable book, Hawthorne deals with both the practical and the philosophical."

~ Elaine Marcus Starkman, author, *Learning to Sit In The Silence: A Journal of Caretaking*, Papier Mache Press

"*Finishing Touches* is a collection of thoughtful, articulate, honest essays reflecting the inward journey of aging – a journey not of self-absorption, but of self-knowledge."

~ Dr. Ann E. Gerike, Ph.D. author, *Old is Not a Four-Letter Word*

Finishing Touches:

An Insightful Look Into the Mirror of Aging

by

Lillian S. Hawthorne, M.A.

ELDER BOOKS

Forest Knolls, California

Copyright © 1998 by Elder Books

Library of Congress Cataloging in Publication Data Main Entry
Under Title:

Finishing Touches: An Insightful Look Into the Mirror of Aging
Hawthorne, Lillian. S.

1. Aging 2. Family relationships

ISBN 0-943873-41-X

LCCN 97-061450

Cover & Book Design: Bonnie Fisk-Hayden
Cover photograph: Judith Less © 1972

Printed in the United States of America

Table of Contents

Part IV PARENTS and CHILDREN: BELOVED ENEMIES

Part V FEELING OUR AGE

Part VI ENDINGS and LOSSES

Introduction

The purpose of this book is to tell what it is like to grow older today — not the pathologies or tragedies of aging, but the expected and inevitable changes that occur. There are changes in how we look and feel and think; there are changes in the friends and family we relate to and how we relate to them; there are changes in what we want to do and what we can do.

We older persons today are not only different from what *we* used to be, we are also different from what older people *before* us used to be. According to current statistics, we are probably the healthiest, most active and "youngest" older generation in history. We are also, according to these same statistics, the most numerous. Our generation of "Gray Boomers" now comprises approximately one-eighth of the total American population and, in another decade, is expected to increase to about one-fifth.

Since there are now so many of us, and we are living so much longer and stronger, we have become the subject of much attention and discussion, especially by a variety of new gero-specialists and "experts." So why write another book on the subject, and what makes this one different from what has already been written?

First of all, this book is different because it is not written about someone *else* who is aging, but by someone who *is*; therefore, it presents a personal, rather than professional, perspective. Secondly, it does not attempt or presume to provide advice or instruction; it attempts only to illuminate. (Indeed, most of us who are becoming elderly are usually realistic enough not to expect solutions, but only hope for understandings.) Thirdly, the aging changes this book describes mirror the experiences and feelings of most of us who are growing older. In short, this is not a "how-to" book about aging, but about "how it is."

If this book can be said to have a central theme, it is that aging is neither the best of times nor the worst of times, but a special time that combines both. It is not a tragedy despite our pains and losses; nor is it a triumph, despite our survival. It is a mixed experience filled with ambivalence and contradictions. We have more time and we have less time; we are more liberated and we are more limited; things come together and things fall apart. If there is a single term that sums up the aging experience, perhaps it would be "bitter-sweet."

I chose the title "Finishing Touches" because it seemed to be both factual and symbolic. Aging is the final stage of life for those who live long enough but, in addition to ending our lives, it adds to our lives, as well. The expression, "finishing touches" refers to those last special efforts without which the entire picture would be less attractive, less meaningful and less complete. These are the "touches" that truly finish, bring out and bring together, all that has gone before.

The material in this book is divided, for purposes of clarity, into seven sections, each representing a major aspect of the aging experience. The sections are not arranged in any order of chronology or importance, but in what seems to be the developmental order of our aging process. Each section contains a number of articles of personal commentary related to that subject; but, unavoidably, as in our own aging, these different aspects sometimes overlap and intersect each other.

The first section, entitled "Changing Faces, Changing Places," looks at such issues as our "new" identity as old people, and our place, both physical and social, in the larger society. The second, "Changing Roles and Goals," considers the different ways we use ourselves and our time, and the different people with whom we use it. The third, "All in the Family," describes our changing relationships with our aged parents, our aging spouses, our mature siblings and our growing grandchildren. The fourth section, "Parents and Children: Beloved Enemies," focuses on the changed and charged relationships with our now- adult children, who seem to combine being friends, foes and family. The fifth section, entitled "Feeling Our Age," describes the health changes we experience, ranging all the way from minor inconveniences to major threats. The sixth section, "Endings and Losses," reflects on the people and possibilities we lose as our lives become more shadowed by mortality. The final section is entitled "Learning Our Lessons," and it tries to sum up our feelings about our aging and the meanings and messages we have been able to find in the process.

My sources for the material in this book are both professional and personal. I have had extensive professional education and experience as a psychiatric social worker, particularly in counseling older people. But more importantly, I am a senior citizen myself, and I have personally experienced, have observed, or have been involved with all of the issues and feelings described in these pages.

Finally, I wish to add a deeply personal note because, for me, writing this book has been like taking "a journey inward." In examining and reporting the experiences of my own aging, I have also been reviewing my life and rediscovering myself. In doing this, I have become more aware than ever of those special people who have been such an important part of this book and of my life. They are: my parents, gone now but always with me; my two daughters, now adults, but always my children; my grand-children, my immediate joy and future hope; and my husband, for better or for worse, but forever. I want to give my special thanks to each of them, not just for making this book possible, but for affording me the pleasure of loving them and the privilege of being loved by them.

Part I

Changing Faces, Changing Places

Through the Looking Glass or, "Who is That Stranger in the Mirror?"

Lately, I've noticed that mirrors aren't made as well as they used to be. The face that is reflected back at me isn't the one I remember. This face has thinly etched lines in the cheeks and forehead and at the corners of the eyes. The mouth is serious and tight, and the hair is thickly striped with gray. There are creases and folds in the neck, like crinkled crepe paper. This picture is in sharp contrast to the one that I see in my *mind*, which is the smooth face of a younger woman with dark hair and a taut, slender neck. Sometimes, when I pass a mirror and accidentally glance into it, I am surprised by the stranger I glimpse there. *Who is that older woman who has taken my place?* When did I change like that? I don't remember noticing it happening.

We don't usually think about growing old, at least not consciously, except in particular situations or in response to particular causes or events. Certainly, becoming old is expected, if we are fortunate to live long enough; nevertheless, we are surprised by it when it does come. We tend to notice changes most readily in people we see least often. But we see *ourselves* all the time and, at least in our *mind's* eye, we see ourselves through the compassionate veil of memory, not the cool, dispassionate vision of a stranger. Therefore, except for the tangible changes we cannot ignore — like glasses we wear, or hearing aids we insert — we tend not to see the changes in ourselves.

As the process of aging progresses, we tend to view ourselves more in terms of how we *feel* than how we *look*. So if, most of the time, I feel capable and active and well,

1

where did that older woman in the mirror come from? Is this how other people now see me? Is this the way I must now perceive myself, not as I *think* I am, or as I remember that I used to be?

It seems to me that getting used to these changes in the way we look may be one of the first and most unsettling tasks of aging. In fact, it may be even more difficult than accepting the changes in the way we feel. At least those aches and pains we suffer, though discomforting, are usually private and invisible. We can hide them in our solitary moments, or only share them with intimate partners, at our own discretion. But the way our faces and bodies look are visible for all to see, despite hair styles or hair colorings; despite skin creams, foundations and powder bases; despite clothes that drape, enfold or disguise; despite all the cosmetic, age-delaying actions we may attempt.

All of us want to live longer, but none of us wants to become old; and especially, we don't want to *look* old. For, in a time and society like ours, in which youth is so prized, growing older is a serious business. Financier Bernard Baruch, who lived until his late eighties, was quoted as saying that old age was 15 years older than whatever he was. We tend to feel a difference between accepting our age and accepting being old, between the way we know we look and the way we still feel. Can those who see us tell that the wrinkles are only on the *outside*, not on the *inside*?

The philosophers, psychologists, and other professionals all tell us that we should learn to accept our aging process. I *do*, I *do* — but that doesn't mean that I always *like* it. George Burns, at the age of 95 or so, said that he didn't think anymore about being young or old, but tried to look as good as he could for the age he was. So why do I feel complimented when I'm told I don't look my age?

I know that I liked it better when I (and others) enjoyed the way I looked in a bathing suit. I liked it better when I could wear sleeveless dresses that showed firm, bare arms. I liked it better when my hair was thick and dark, and when my husband's hair was still *there*. I liked it better when stairs weren't so steep, when packages weren't so heavy, and when walking distances were a more manageable length. And I don't like it very much now, when I request a senior citizen discount at a store or theater, and no one even questions it.

That stranger in the mirror takes some getting used to. I recognize the resemblance to the face I remember, (like a familiar scene showing underneath a translucent overlay). I accept this different person as myself with some feelings of surprise, regret, and resignation. I remember the vanished younger person underneath with affection, nostalgia, and some momentary grief. And I try to accept both of these selves, so as not to invalidate either one. However, deep down, I know that I liked it better when mirrors made me want to smile, rather than laugh or cry!

Junior Seniors and Senior Seniors

We don't seem to know exactly when old age begins, because there are so many different definitions and timetables, depending on which biological, psychological, or sociological clock we use. According to Medicare, we are old at 65; according to Social Security, we are old at 62, 65, or 70, depending on when we choose to collect, or how much we can earn. According to certain organizations, we are not old until 70, which is their age for mandatory retirement. According to various retirement communities, old age begins at 55, which is their age for senior residence; according to certain businesses, senior citizenhood may begin at 50 which is their minimum age for special elder discounts. According to physiological indicators, old age may not start until 75, because we are in so much better health than previous generations.

But even though the calendar or chronology of aging may vary, we do know, as older people, that there are more and more of us around, and we seem to be around for a longer and longer time. At the beginning of this century, only one person in 15 was over age 65; by the end of this century, it will be almost one out of every four! In fact, not only are more people living longer, but we are also growing older more slowly, so that the "old" are now older than they used to be.

Just in this past century, our life span has been extended by almost 30 years, which is virtually equivalent to an entire added generation. These extra years have not merely been attached at the very end of our lives, but seem to be inserted somewhere in the middle, so that we remain middle-aged longer functionally, even though not chronologically. In other words, longevity has not just added more years to our lives, but "more life to our years."

It used to be that aging took up only the last few years, and only a small proportion, of our lives; but now, we spend more years being older than we spent in being children, adolescents, or adults. We may remain senior citizens for as much as one-third of our entire lives, so that we now have more graying years than growing years. As a result, we are different at the beginning of the aging period than we are at the end; it may even be hard to believe, at times, that both are part of the same life stage. Our aging now tends to occur in two stages: first, as "junior seniors" in our 60s and 70s; and later, as "senior seniors" in our 80s, or even 90s. At the beginning, we seem more like our own (almost middle-aged) adult children; at the end, we become the way we finally remember our parents.

As "junior seniors" — especially at the beginning — we may still look, feel and behave like those in their middle age, and we probably feel that we have more in common with them than with the really elderly of our own generation. Although we may all be seniors, some of us are more senior than others; and even if the young may lump us all together in the same age group, we ourselves do not. We are still physically active and physically attractive; we are still playing tennis (even if it's now doubles instead of singles); we are still dancing, driving, and daring.

In the retirement community in which I live, residents range in age from their early 60s to their late 90s. The median age is late 70s, and those who "aren't even 70 yet" are the "younger generation." When we recall our parents at this age, we know that they were older, slower, different than we are. If we are reasonably healthy and reasonably financially secure, we are, in almost every respect except

actual chronology, about 10 years younger than our coun-
terparts of previous generations. In fact, being a "junior
senior" is not at all an unpleasant stage of life, and we
wouldn't mind getting older, if we could remain this way.
Although we know we are too old to be young, we feel too
young to be old!

Sooner or later, if we live long enough — regardless of
what or where the age boundaries are, regardless of how
well we feel, or look, or act — we move from being "junior
seniors" to being "senior seniors." In the earlier stage, it
was still possible — because we were still so "youthful,"
and also because there were so many others who were so
much visibly older — to deny or ignore our old age. When
we become "senior seniors," this is no longer possible.

As we age, we find there are more and more things we
can no longer do as well, or as much, or cannot do at all by
ourselves — even things we once did for others. We now
walk more slowly and tire more quickly; we sleep less, but
doze off more; our bodies do not move around much, but
our minds wander a great deal; we find that there are some
things we cannot remember at all, but some things we cannot forget. There are liver spots on our skin, cataracts in
our eyes, arthritis in our joints, and what sometimes feels
like cobwebs in our brains. Our minds and our bodies
grow finally, and fully, old.

Living in a retirement community, as I do, "senior
seniors" are my neighbors and friends, as well as my
future. Although those of us at the beginning of the aging
process may not now recognize ourselves in those at the
end, we recognize what we will become. On the one hand,
it may be discomforting to see what our future selves will
be; but on the other hand, it can also be comforting to see
how much and how long life can still last in our "older"
selves. In other words, "senior seniors" can not only pro-

vide us with an introduction to our future, but also with an inspiration for our future.

The Feminization of Aging, or "The Woman's World"

When we grow older today, it virtually means that we move into a woman's world. We know that women outlive men by several years; more than half of the older women in America today are widows, whereas less than 10% of older men survive their spouses. Not only are there more females over age 65 but, as we move up the age scale, the proportion increases, so that the over-80 generation is now actually two-thirds female. Middle age has been described as a period ruled by powerful men, and old age as a world inhabited by surviving women.

Historically and traditionally, this would not have been expected. First of all, the stereotype of being "the weaker sex" would have predicted greater, not lesser, mortality. Secondly, given their traditional upbringing, women, if they survived at all, should not have survived so well.

Women in our generation, unlike our adult daughters or growing granddaughters, were not prepared, nor did we expect, to live independently. We went directly from the parental home to the marital home. We generally defined ourselves in relation to the important men in our lives, i.e., our fathers, then our husbands, and finally, even our sons. While the men were the achievers and providers who were engaged in the outer world, women were traditionally encased in the inner world of home and family. It is not until the later years, when women find themselves surviving alone, that they also find themselves, for the first time, in charge of their own lives.

Yet today's generation of older women is coping unexpectedly well and seems to be able to draw upon special, unsuspected reserves. It seems to me that their survival strengths have derived from three factors: 1) their socialization skills; 2) their support systems, and 3) their role stability. Ironically, each of these evolved from the very traditional circumstances in which they were raised.

First of all, since women were typically assigned personal rather than professional roles, it became their task to deal with people, to extend hospitality, and to develop social relationships. It is these learned social skills that stand them in good stead now, as they relate to people on their own.

Secondly, since women were generally home-based, they developed associations with other women at home who shared their daily experiences and needs. These women exchanged practical advice, encouragement and understanding. Over the years, as chronology altered life experiences, the activities of these groups changed also, but their mutual support continued, unchanged by time.

Finally, since women have traditionally been the family caretakers — starting with young children, continuing with ailing or failing spouses, and finally, with elderly parents — they have always had an important and recognized lifetime role. In fact, it has been estimated that the average woman can now expect to spend as many years taking care of her aging spouse or parents as she did in childcare. On the one hand, this lifelong task is a burden; but on the other hand, it may also be a bonus, because it provides a continuing sense of purpose and possible source of satisfaction.

This is in contrast to older men, for whom retirement brings such drastic changes in activities and associations. Feminist author Betty Friedan and psychoanalyst Erik

Erikson agree that women probably handle aging better than men, because growing older brings them fewer feelings of loss and more of opportunity.

Indeed, in a recent study comparing male and female longevity, researchers concluded that, because men's self-esteem was so attached to their sense of strength and work achievements, growing older was more difficult and, therefore, affected their health and their mortality. In other words, for women, "weaker" turned out to be "wiser;" the very coping skills they developed as a result of their traditional circumstances enabled them not only to survive longer, but stronger.

Although women do so well as survivors, this does not minimize the loneliness of their longevity. What they suffer is actually a dual loneliness. First and foremost is the particular loss of the special person in their lives — the spouse with whom they have most likely lived most of their years.

But there is also a feeling of general loss, of missing not men's *power*, but their *presence*, in their lives. Living in a uni-sex society, without masculine sounds, styles, and stimulations, is somehow tame and drab. Who is there to argue politics with, or discuss sports scores with; to compete or conspire with at the bridge table or tennis court? Who is there to consult with, contrast with, or complain about? The color, conversation, and even controversy that men bring are missing from the women's world. Certainly women don't regret surviving; after all, as Eda LeShan states in the title of her newest book, "It's Better to Be Over the Hill than Under It!" But women miss what they have lost; they miss the men in their lives.

Olden Years and Golden Years

The difference between the "olden years" and the "golden years" is only one letter, but perhaps it really should be a sign instead — a *dollar* sign! Money is our biggest concern, second only to health, when we grow older. Health has been called the blessing money can't buy, but we know very well that there aren't many other blessings we can afford without money. Having enough money is certainly important at all stages of our lives; but when we're young, there's still time and hope for financial fortune. When we're older, we realize that what we have now is all we *will* have.

It's not that we weren't aware of, or concerned about, financial security in our younger years. Most of us did some advance planning because retirement is not an unexpected event. In fact, as children of the Depression, we were very much imbued with the work ethic and with cautious patterns of spending and saving. We were worriers about the future; we worked hard; we lived prudently; we postponed luxuries so that we could afford a comfortable and secure old age. We were raised to save for "a rainy day," but the weather has turned out to be different than we expected; the rainy day has become a long, slow drizzle!

The major change is that we're living much longer, which is both a pleasure and a plight. It means that our fixed financial resources now have to be stretched out over a longer period of time. Our money must be able to provide not only for quality of life in our early retirement years, but also for the special needs to sustain life in our later years.

It also means that inflation has more time to shrink the purchasing power of our dollars.

Our attitude toward money changes when we get older. We're concerned more about security than success, about protection, rather than profit. But how can we tell in advance how much will be enough to ensure a trouble-free old age? There's no way to guard against creeping inflation and the increased cost of living. There's no way to guard against increasing needs for medical care and increasing costs of that care. We cannot just assume that our savings and subsidies will automatically be able to sustain our life style. Instead, we have to continually eliminate and adjust, in order to leave enough allowance for the abundance of years and the inroads of the unexpected. Getting old is unavoidable if we live long enough, but being poor should not be.

It's true that certain expenses diminish as we grow older. We probably spend less on food, clothing, and entertainment; we're less concerned about acquiring material things; our home may be all paid for by this time (and also worth a great deal more than we paid for it originally). But these advantages aren't real ones. Where would we live if we sold our homes, even at a large profit? What do we do with the money we save by giving up things that may make life pleasurable? We hoard it, so that we can later afford things that life makes necessary.

Not only do we need enough money to age comfortably, but also to have smooth relationships with our families. One of the pleasures we want to be able to afford is generosity to our children and, more especially, to our grandchildren. As a matter of fact, sometimes we need to provide financial assistance to our adult children for new homes or new businesses, or for our growing grandchildren's college educations. We joke sometimes about spending our children's inheritance, but that's really more fearful than funny. One of our greatest dreads, as we grow older, is to become financially dependent on our children; one of our most cherished hopes is to be able to leave our children a tangible legacy.

Speaking about our children, I wonder sometimes how they will manage their own financial futures when the time comes. Actually, the time is not so far off, because our adult children's generation is entering, or already in, their 40s, and beginning to think about retirement planning. It seems to me they will have both an easier and a harder time. On one hand, they're probably more knowledgeable and sophisticated than we were about investments and money management. But on the other hand, they have a different lifestyle, with less financial waiting or worrying. In one of her recent books, Erma Bombeck wrote that our generation's idea of the American Dream was " ... 20-year house payments, 40-year marriages, and 30 years on the same job." But our children have a different timetable. I wish we could give them wise advice learned from our own experience, but how can we help them anticipate the unanticipated? Besides, they probably wouldn't listen, anyway!

Overall, my husband and I feel lucky because we have each other and each other's financial support. We're grateful that we're as financially secure as we are, and we feel better off than we remember our own parents being at this time in their lives. We feel comfortable right now, in our 70s, but we're still worried about our 80s.

It seems that our generation prepared for the expected, but the unexpected occurred — longer lives, higher prices, catastrophic medical costs. We find that we have not one, but several old ages, and each of them brings different, and more costly, needs. I know we all want to have more gold in our "golden years" but, like developing any other precious substance, it seems to require some learning, a lot of labor, and also, a lot of luck!

The Rest of Our Years, the Best of Our Years

The society in which we live now defines not only *when* we grow old, by its policies toward aging; it also influences *how* we grow old, by its *attitudes* toward aging. The problem is that society sends us confusing, and sometimes contradictory, messages about being older.

First of all, there is the matter of physical appearance. On the one hand, white hair and weathered faces were traditionally supposed to reflect wisdom and authority, and to earn us respect. However, the media, cosmetics, and fashion industries make it clear that desirable physical features belong only to the young. The accepted role models of beauty display looks and shapes that are in stark contrast to our own aging bodies, faces, and forms. We find ourselves inundated with the advertised promises of cosmetic surgery, hair-colorings, anti-wrinkle creams, diets, pills, and special tonics to make us young, or keep us young or, at least, help us *look* young. We are sometimes tempted by these promises because, no matter how well we may feel or function, it is hard to ignore the message that to be young is to be more attractive and more desirable than to be old.

Then there are the matters of economic and personal independence. When we were young, we were advised to save our money, spend carefully and invest wisely, so that we could be financially secure and independent when we were old. We followed that advice but, today, because we were prudent and fortunate, we are castigated as "greedy geezers" who are enjoying too many financial benefits which younger generations cannot afford.

Society also encouraged us, as it did *all* citizens, to be self-assertive and active on our own behalves, rather than become dependent on others. So we became involved in community organizations, advocacy groups, and senior citizens lobbies. Yet we are often criticized for trying to exercise too much influence, of being too politically involved, or even of voting too much!

Finally, there is the matter of our longevity. Society has given us the tools for staying healthy longer and living longer. We now have improved nutrition and sanitation, and more modern pharmacology and medical technology. But, in subtle (and sometimes, not-so-subtle ways), we are being made to feel guilty for living so much longer. We are repeatedly reminded that, because of our longer lives, we are using up disproportionate amounts of society's financial, social, and health resources. The longevity we now enjoy is not something we sought or expected, but it is a gift we cannot refuse — nor will succeeding generations, when it is offered to them.

In a sense, some of this confusion and contradiction is understandable, because the increasing number of older people is a new human phenomenon. In the past, old age was rare and didn't last long, but today, the old are getting both older and more numerous. It is predicted that by the middle of the next century, the number of senior citizens will be double the number today. No wonder society really does not know what to do about, or with, so many of us older people.

Although aging is a process that eventually affects all of us, if we live long enough, many of us really do not know much about it; nor do we always want to know, or even think much about it. Perhaps this is because growing older reminds us of, and brings us closer to, our own mortality. As a result, many myths and stereotypes have arisen about age, virtually all of them negative. Older people are pictured as frail and failing, anxious and unhappy, insular and incompetent. They are thought to be bored, or boring, or both; too rigid to change or learn, and too self-centered to care about other people or other problems.

The truth is that gray hair doesn't necessarily mean a gray outlook, and researchers on aging have actually come up with very different pictures of the elderly. It turns out that people grow less anxious or fearful and less sensitive to pain when they are older, due to both psychological and physiological causes. Elderly interests and curiosities can actually expand with age, even though some physical capabilities may contract. Older people become less conformist and more individualistic, so there are more differences among the elderly than among the young (perhaps because there have been more years to *become* different). Overall, it was found that older people are happier than young people think — probably because, being young, they cannot comprehend anyone being happy being old.

Yet, even though we know that no one who lives long enough can avoid it, we are unprepared for becoming older ourselves. I know that other people are getting old; I know that my parents were old; but somehow, it seems surprising that it is actually happening to *me*. I remember that the first time someone referred to me as "a woman of your age," I had to look around to see who that person meant. Then, with surprise and embarrassment, I realized it was me!

When Eda LeShan warned that "... being over the hill is better than being under it," she may not have realized that this attitude makes aging no more than an unwelcome, though necessary, evil. I realize that no one looks forward to growing older, yet aging is really neither a sign of defeat nor a punishment. In fact, it has been observed that, "If age is no good, then life is no good ... because that's what it leads up to." It has also been said that "the winter of life can be an education in itself because, even though *everyone* is too old for *something*, no one is too old for *everything*." Therefore, aging can be a time of growing, as well as graying; what remains can be not just the rest of our years, but can even contain some of the *best* of our years.

Long-Time Friends,
Changed And Unchanged

Over the years, we have many different friends, and also different *kinds* of friends, at different stages of our lives. In our younger years, our friendships are usually more intense and intimate, as we share with one another our innermost thoughts and our dreams of the future. In our friendships in later years, we are more interested in sharing support than self-revelation, and we tend to expect less, and give less, to each other. In fact, when we are older, we rarely make new close friends anymore. We may have less energy now to invest in new intimacies; or we may have redefined, in a more limited way, our expectations of one another; or we have no shared history with each other, and know there will never be time to build one.

Psychologist Judith Viorst defines several different categories of friends during the course of our lifetimes: There are "convenience" friends, people with whom, or for whom, we do occasional specific things, or share certain particular activities. There are also "crossroads" friends, people with whom we once shared a special, time-limited experience or event in our lives. And then there are "historical" friends, people who knew us "when," and were an ongoing part of our past lives, but who have now gone their separate ways.

Finally, there may be a few special, close, "forever" friends. If we are lucky or if circumstances permit, or if we make sufficient effort, we may be able to retain such relationships despite time and distance. My husband and I are fortunate to still have a few, special, long-time friends whom we see regularly at designated intervals, or on special occasions; even though we are geographically apart, we have not grown personally apart.

Each time we are together, I am struck anew by how much each of us has changed since the last time we met, as well as by how much we have remained the same. What I notice first of all are the changes in the way we look. Even though we carefully keep ourselves slim, each year, the pounds are more visible. Even though we keep ourselves artfully groomed, each year, the lines in our faces are deeper. Even though we take reasonable care of our health, each year, we are a little more vulnerable.

Yet the ultimate compliment we pay one another when we meet is, "You haven't changed a bit!" Aside from the fact that it is most likely not true, it also sends the message that the way we used to look is preferable to the way we look now, and that looking younger is better than looking older. Perhaps what we really mean to say to each other is that we are still the same people we were; it is only our faces, or our bodies, that are different.

When I see my women friends, I know that the changes I see in them mirror those I cannot, or will not, see in myself. And when I look at their husbands, I can see in those men aging changes I do not notice in my own husband (whom I view with the eye of memory and affection, as much as with the eye of reality). Even though we cannot see in ourselves, or in those closest to us, the changes we see in others, we know that others can see those changes in us. So when I meet with my long-time friends, I wonder about what they see when they look at me. Can they still see the young friend I used to be when our friendship first began, or only the old friend I have now become?

Yet despite the years, miles or changes, enough has remained so that when we meet, even only periodically, we know that we are not strangers to each other. In fact, I seem to view these long-time friends through a kind of double exposure in which the present and the past are superimposed upon each other. When I look at them, I see two sets of images; when I speak with them, I hear the sounds of two voices; at the same time, I experience the person I remember and the person I now perceive.

I feel connected to them by the experiences, relationships, and memories we have in common. In fact, when we are together, we are able to speak a kind of shorthand language which is really the special vocabulary of shared history. When one of us mentions a certain name or place or incident from the past, the other immediately remembers and understands, without need for further explanation. Even if our recollections are not totally accurate, or do not absolutely coincide, the differences in details are less important than the fact that we are connected to a common past.

Being able to share memories with old friends is important because, with our parents gone, our siblings possibly scattered, and our recent friends knowing only our present selves, there are few people left in our lives who are links with our past. It is not that we want to relive the past, but we don't want to lose it; and it is lonely trying to remember it by ourselves.

In addition to sharing experiences and memories, we also share similar lifestyles and life histories. We hold many of the same values and make many of the same assumptions; we seek many of the same goals and avoid many of the same pitfalls; we enjoy a similar sense of humor, and are moved by many of the same sorrows; we feel similar worries, complaints, and even fears. We share what has been called "a common point of departure for our journey through life." Indeed, in some ways, we actually feel more alike now than when we were younger, perhaps because we feel our sense of generational kinship more strongly at this time in our lives. I suppose that is why, even though we recognize in each other what is different, we can still relate to what is the same.

Retirement Communities: Garden of Eden or Gilded Cage?

More and more of us, as we move into senior age, are also moving into retirement communities. And at this stage of our lives, that's what we're looking for — community, as well as retirement. Most of these communities have euphemistic names which include words like "golden" or "sunset" or "leisure," but everyone understands what those words really mean.

Before World War II, retirement communities hardly existed; now there are more than 2000 of them, and nearly one million of us living in them. When my husband and I retired and moved into a retirement community a few years ago, we expected to be living a different lifestyle in a different life setting. But we weren't completely prepared for all the transitions and trade-offs we found there.

The first thing that is so striking, because it's so visible, is the age segregation. We are surrounded by the aged; youthful people and children are only briefly and occasionally present. My husband once remarked ruefully about the bikinied sights he used to enjoy at poolsides; now he must await the nubile granddaughters who come to visit and be viewed! But for me, it's actually comforting to see all the other women around me looking like I do in a bathing suit. So, although I miss the pleasure of the presence of the young, I know that at this stage of my life, they are not really a part of my life.

We also need to keep in mind that within the age segregation, there is also great differentiation. Our outside shapes and sizes may be similar, but our inside selves are quite different. In fact, we may be even more different from one another than younger age groups, because we've had so much more time to *become* different.

Since we live in an age-segregated community, illness and death are more pervasive than they would be in an ordinary neighborhood. The ambulance wail and the emergency response are common occurrences. That's not pleasant, but it's a fact of aging life. To compensate, we have the concern and assistance of neighbors and groups who keep watch and who help with all kinds of thoughtful, unasked-for favors. So, even if mortality is an ever-present, though uninvited, guest in our community, at least we know we don't have to confront it without help.

All of us here are "empty-nesters," and for those whose own families are far away, neighbors become a surrogate extended family. However, as happens sometimes with family members, they may want to know not just *how* we're doing, but *what* we're doing; they may become invasive, rather than merely involved.

Another thing I notice here is the formidable array of activities. Everyone is hiking, biking, bowling, swimming, golfing, sculpting, painting, folk dancing, square dancing, tap dancing, reciting, and/or performing. Calendars are as filled up as a popular Southern belle's. Although there isn't actually pressure to participate, there is certainly expectation; non-conformity is acceptable, but unusual and surprising. On the other hand, taking part in group activities is a safe and simple way to exercise our minds and muscles. It's a way to meet with people and be with people. After all, at our age, we no longer find life-term friends, but "doing-things-together" acquaintances.

Retirement communities have been disparaged as "vast adult playgrounds." But underneath the games we play, there are other things happening — namely, aging, illness, worrying, and courage. So perhaps it's not surprising that we feel the need to do as much as we can while we can. Residents of retirement communities have been described as being in an "endless autumn." But isn't that better than entering an early winter?

An important reason many of us moved into retirement communities is security. We were seeking to escape from urban crowding and crime, from noises and nuisances, to rediscover a quality of life lost during our busy years. It's ironic that when I first found myself in this new security and serenity, I had to be re-educated *not* to lock my car and not to walk alone without fear.

There's a wall that surrounds our retirement community; it's not high enough to keep out a determined intruder, but high enough to keep out the outside world. It also keeps us inside. I worry sometimes that it may be too safe here, that I may become too separate from the "real" world outside. I venture forth as often as possible, and I enjoy the diversity and excitement outside these walls when I do. But more and more, I feel relieved that these ventures are only visits, and that I can return home when they're over.

I realize that the amenities, security, and support systems here all have a price tag — in trade-offs, if not in currency. I also realize that wherever we move, we take ourselves along as our own interior decorators — of our lives, not just our homes. I've found that the retirement community is not the Garden of Eden — nothing ever really was, or can be. But it's also not a gilded cage, unless we construct our own bars.

Far From the Old Folks at Home

In past generations, families lived in the same homes, or at least in the same neighborhoods, all of their lives; and when their children grew up to have families and homes of their own, they lived nearby. In contrast, today, one out of every two Americans does not even live in the same *state* in which he or she was born, let alone in the same community.

It is ironic that, in the past, when older parents did not live as long, they and their adult children lived with, or near, each other, in extended families. But now that we are living longer, we are living further apart, precisely during those years when we have the greatest need to be closer together.

I am one of those older parents who lives far away from my grown children. For now, I enjoy my independence and my independent relationship with them. For now, we speak to each other often on the phone; we visit when we can; I know that if there were a major crisis, they would respond quickly, although they could not stay long, because their lives are located elsewhere. For now, as long as my husband and I are together and in reasonably good health, we feel safe and satisfied, despite the distance from our children.

But what if either or both of us should become seriously ill or incapacitated? What if we should need help, not just in a crisis, but with continued assistance? What if only one of us remains, as inevitably must occur; how does the survivor manage alone?

I have begun to think about such things more and more

lately, and I have begun to notice, with both sympathy and alarm, what has been happening to some of my acquaintances in the retirement community in which I live. Many of us who do not have relatives close by probably chose to move into a retirement community precisely because it could provide a substitute family; but the time may come when that is no longer sufficient.

Those of us who do have children nearby have begun turning to them more and more for advice, assistance, and support. Those whose children are far away — especially if they are widows or widowers — have begun to make the decision, or to have the decision made for them, to relocate closer to where their children live. They are the ones who must move, since they know that their children cannot change their own lives or homes to be nearer to them. They usually explain the move either as a response to their children's concern and invitation, or as an expression of their own concern to spare the cost and time of long-distance visits. Either way, they can present it as an act of parental assistance to their children, not the other way around; either way, they can present it as a voluntary choice, whether it is or not.

Moving nearer to one's children can *raise* problems, as well as solve them. Our adult children are now at a point of maximum responsibility in their personal and work lives, with their careers and businesses approaching finality, their growing children approaching costly college years, and their own approaching middle age. Their lives are busy and far different than ours. We do not want to become another burden for them when we know they already have so many; nor do we want to reverse roles and become their dependents instead of their parents. Also, we may not fit into our adult children's lifestyles, schedules, or neighborhoods. So, we could find ourselves just as alone near them as we are when far away.

I know that I have envied parents whose children are nearby, or at least near enough to be comfortingly available. I also know that this cannot be the case in my life, unless I am willing to be relocated and uprooted. And I know that, realistically, I will be unable to turn to my friends when the time comes, because they are my peers and, by the time *I* need help, so will *they*. I do know that there are professional resources available in my community. But, as difficult as it is to admit the need for help from a loved one, it is even more difficult to contemplate the need for help from strangers.

I remember, several years ago, hearing my elderly father reminisce about how it used to be in his parents' time, how aging parents lived with their children, especially their daughters. I remembered that I responded, in self-defense, that daughters now had careers of their own, and that different generations led different lives. All of this was true, but I still felt guilty for having said it. Now I can understand better what my father was feeling, and I realize that we were both right. He was not really being critical of his own children, who did not, and could not, take care of him in the same way as was done in past generations. He was simply stating truthfully that that was the way things were in a different time of the world, and in a different way of life.

Now I am the older parent, and I am the one who wonders and worries about the possible need for care from adult children. Actually, it is not so much wanting to *live* close to them, but wanting to *feel* close to them. If I did not know that, realistically, I might some day need some kind of help, the distance between us would not matter; and if I did not know that my children genuinely want to be able to help, the distance also would not matter. The dilemma is that we care about our independent lives, and we also care about each other. As parents, we really do not want to have to ask our children for help, but we fear that we may have to; and as children, they do really want to help us but, because they are so far away, they fear that they cannot.

Part II

Changing Roles and Goals

Retirement, or "I Married You for Better or for Worse, But Not for Lunch!"

It's been said that one trouble with retirement is that there's no future in it! When we retire, we're usually given some kind of present, some kind of party, and some kind of praise; but we also receive one more thing, a new and public identity as a "senior citizen." Before retirement, no matter what our age, our identity is defined by our work; after retirement, no matter what the work, our identity is defined by our age. Retirement is a rite of passage, like graduation, to let us — and everyone else — know that we must now face the reality of our own aging.

Those of us who retired recently can expect to spend almost as many non-working years as we did working; that amounts to about one-third of our lives. Our grandparents, and even our parents, mostly worked, and then died; but we're retiring younger now, and living longer. The irony is that now that we have so much time without working (which is what we dreamed about all the time we were working), we haven't quite figured out what to do with it. One problem with retirement is that we never get a chance to practice in advance!

I remember when my husband retired, he commented that although he didn't miss the suit and tie, he did miss the office copying machine. I suppose that's another way of acknowledging that retirement brings both gains and losses; in fact, sometimes the same change of circumstance can be both a gift and a problem.

For instance, consider the matter of time. We're now freed from the alarm clock and the calendar, giving us a long time with a lot of time to use, or even lose. But I still feel guilty if I'm not being busy and beneficial; and the abundance of time sometimes seems like an emptiness that I must keep filling up.

Then there's freedom from responsibility. There are no more job assignments or expectations. But now I feel strange — both comforted and disappointed — to learn that the job tasks are being accomplished, and the goals achieved, without me. When I talk to former colleagues, increasingly briefly; and when I visit the office, increasingly infrequently, I notice that there are fewer people who remember me. The person sitting in my place, at my desk, in my office, looks very comfortable there; I'm the one who's now the stranger in the room.

There's also freedom of choice. No longer do job demands decide for me what I should do, when I should do it or with whom. Now that I can make my own choices, I feel exhilarated, and also unnerved. There's suddenly such an array of possibilities, and so few directions or deadlines. Now I have to make all the effort, take all the risks, and shoulder all the responsibility. Can there be such a thing as an "overdose" of freedoms?

When my husband and I first retired, we reacted to this new abundance of freedoms in opposite ways. I chose to "overdose" on activities — taking courses, joining organizations, doing volunteer work — causing my husband to wonder when I'd ever had time to work. He, on the other hand, preferred the opposite path of mandatory leisure; but the disadvantage to that path is that it takes all the novelty out of weekends!

I've concluded that men and women retire in different ways. Most men retire once, and all at once, and it's a major culminating event after a lifetime job or career. Women, on the other hand, tend to retire several times, from different roles, i.e., from parenting when our children grow up and leave home; from partnership in our husbands' work lives (even though we get no retirement parties for these), as well as from our own jobs, if we worked outside the home.

Not only do we and our husbands retire in different ways, we also have different expectations of retirement (that don't always match!). A woman's fantasy is to go out more; a man's is staying home more. Now that there's no more fixed work schedule, I look forward to dining out and traveling far; now that there's no more fixed work schedule, my husband looks forward to home-cooked food and Monday night football.

It's ironic that we married a long time ago in order to be *together*, and ever since, found ourselves separated by our jobs, our individual responsibilities, and our individual activities. Now, finally, in retirement, we find ourselves together after we've grown accustomed to being separate. Now we're suddenly expected to spend 80% of our time together, after we've been spending approximately 70% of our time apart. Having a full-time, live-in spouse, for the first time after all these years, takes getting used to for both of us. As my husband is at home more, he shares more in household activities. And I have to learn to allow him to share, with appreciation for his effort, rather than resentment at his intrusion.

In a way, it's almost like when we were first married; we have to redefine our roles with each other. This takes both common sense and uncommon sensitivity, plus the healing help of humor. The poet Robert Browning once invited his love to "Grow old along with me. The best is yet to be." It's an invitation I extend to my husband, too, in our retirement; though I still struggle, at times, with the thought that, "Yes, it's for better or for worse; but, please, not for lunch!"

Change of Life and Change of Lifestyle

Growing older brings changes in our lives, as well as in our *lifestyles.* Aging can offer us a second chance, almost a second childhood. This may be our last chance to become whatever we wanted to be when we grew up (which we were too busy to become while we were actually growing up).

For one thing, we didn't have the time; we were too occupied with the work of day-to-day adult living. We were too busy in our offices, stores, factories, laboratories, and classrooms. We were too busy in our homes, PTA meetings, scout troops, and carpools.

In a sense, we now actually have both more time and less time. We have more *present* time to do things, but also a foreshortening of *future* time. As a result, we are both more selective and more expansive about what we do. We want to use our time as fully as possible, but also as favorably as possible. By that, I don't mean that we just want to play, but we do want to have pleasure, keeping in mind that time for recreation can also mean time for re-creation. We certainly don't want the years left to us to become leftover years.

This is a time of our lives when we are not so pressed by obligations, by the needs, or "shoulds," of others. Until now, we did what was required or expected of us by parents, teachers, employers, spouses, children, or even neighbors. Now, we can have newer and freer responsibility for ourselves. Now, we can finally let go of the poses and conformities which were required in our previous roles, and we can try on new styles and different postures.

Although our parental and work responsibilities may be finished, *we're not*; we only retired from *work*, not from *life*! We can now say "yes" — or, at least, an interested "maybe" — to opportunities that, in earlier years, were not considered suitable or realistic or available. You see, one of the advantages of growing older is that, instead of having to *find* ourselves, we can just *be* ourselves.

We older people today are healthier, more active and more interested than previous older generations. We remain middle-aged for a longer time and are old for less time, which means that we have been given a gift of additional years and energy. Aging also brings us other gifts: intelligence and emotions that still work; the ability to distinguish between what's important and what isn't anymore; and a continuing capacity to discover new sources of delight. Our joints may become stiff and creaking, but our minds and spirits don't have to.

One of the freeing changes that happens as we grow older is the emancipation from certain inhibitions and self- or other- imposed expectations. We're now in a position to dare more than we ever did in the past, because we're less afraid of failure or foolishness. Now that society pays less attention to what we do, we're free to pay more attention, to become more ourselves (or the selves we never had the time or opportunity to be). We don't have to be flaming successes at whatever we undertake now, as long as we have experiences that engage us and projects that please us. I recently read about one woman in her 90s who remarked that she had so many projects that she didn't know when she would have *time* to die, let alone become sad or sick!

As a matter of fact, I've noticed that, as we do become involved in new activities, we may find that these are not changes at all, but rediscoveries. We may find that old talents and creativities from the past (that were suppressed or stored away or forgotten during the process of our adulthood) have not vanished at all, but now re-emerge, when tenderly coaxed, in this new season of our lives.

For example, my husband recently began taking violin lessons again for the first time since his boyhood. When his young teacher, attempting to be solicitous of this new older pupil, inquired when he'd had his last violin lesson, he blithely replied, "Some time in 1933!" I don't know how the teacher feels, but I know that my husband has been busily enjoying his music.

I have met others who have been painting, acting, sculpting, and writing poetry with some talent, a great deal of enthusiasm, and no embarrassment. Like them, as I grow older, I worry less about failing at what I try to do than about failing to try at all.

We also need to remember that these second chances and new beginnings don't automatically occur. We need to make efforts to find them and foster them. The humorist-philosopher, Sam Levenson, once observed that, "If the best is yet to come, it will only come if invited." The result of refusing or ignoring such an invitation is to become both bored and boring.

Growing older can bring *inner* as well as *outer* changes; it can open new doors, as well as close old ones. It's certainly true that our energy is less and our time span is less, but how we choose to use that time and energy can decide the quality of the rest of our lives. It was Thomas Carlyle, the English essayist, who said, "The tragedy of life is not what men suffer, but rather, what they miss." That's why I have concluded that I would rather risk than regret, that I would rather "wear out than rust out."

Making Friends and Enemies in Our Later Years

When we were younger, we were concerned about making good impressions and good friends. Now that we are older, we are less concerned about how others evaluate us, and more about what we value in our own lives. In other words, our attitude toward making friends and enemies has changed.

We become both more tolerant and less tolerant, more discriminating, but less discriminatory. By that, I mean that we are now willing to accept more differences in other people, but less willing to allow those differences to affect or perturb our lives. We are more patient with the needs and actions of others, but less patient with their demands on us.

Although it seems like a contradiction, we become both more certain and more uncertain than we used to be when we were younger. Psychologist-poet Judith Viorst describes this change in a poem about getting older:

"We aren't as uncertain as we used to be.
We've learned to tell the real from the tinsel and the fluff.
We choose our own thoughts and musts and got-to's and shoulds,
(But) we aren't as judgmental as we used to be.
We're quicker to laugh, and not as eager to blame."

Perhaps we have finally learned to accept the fact of human imperfectibility, and therefore, no longer need, in the words of Judith Viorst "to love and loathe with equal strength." Perhaps, as we become less controlled by the opinions of others, we are more able to be committed to our own inner directions. We learn to separate the needs of others from our own needs, and thereby to accept as well their styles and solutions — which are not necessarily ours. In any case, we seem to become both more accepting and more selecting of others as we grow older.

I think another reason we become more selective about our activities and associations is that we become more aware of our own diminishing energy. Friendship requires energy — physical energy to engage in activities; and emotional energy to engage in relationships. As our energy supply becomes less, and therefore more precious, we do not want to waste our waning resources on people or practices we no longer care about, or responsibilities we no longer need, or excitements we no longer enjoy. This is not selfishness, but self-protection.

As we shed non-essentials, our attachments become limited to the most significant people in our lives. Extraneous persons cease being friends; they do not become our enemies, but merely passers-by in our lives. The shrinking circle that remains consists of our family, which probably becomes our primary involvement, plus special, selected friends. This circle of caring may be smaller, but the attachments are deeper.

There are some observers who see only sadness in this process. Psychologist Eda LeShan calls it one way of "learning to live with loss." Others use the term "disengagement," describing a cut-off of emotional involvement with others. I disagree with these gloom-and-doom descriptions; I see it as perceiving "the glass to be half-full, rather than half-empty." I view this process as rearrangement, not disengagement. We are not eliminating our involvements, but redefining them. Indeed, what we are losing is what is no longer needed or wanted. We do not uncommit ourselves, but disencumber ourselves by accepting the fact that, as we change over time, so do our friendship needs and capacities.

Another thing I have noticed as we grow older is that we have less need to be safely neutral or nice. We become more direct and candid, and less concerned about the impressions, opinions or approval of others. I think this is particularly true of the "old old." They seem to see things without clutter and say things without constraints of convention. Indeed, I have heard the very elderly express opinions — personal, political, and philosophical — that from others would be considered eccentric but, from them, are accepted with equanimity. In many ways, they say things that we may think, but are still too inhibited to express ourselves.

My father died recently at the age of 96, and he was of sound mind and wit almost to the very end of his life. I remember his telling me that one of the advantages of the very old, as also of the very young, is that they are permitted to say and do things not usually acceptable of others. On the one hand, there may be something patronizing about indulging the very elderly as one would young children. But, on the other hand, it also offers a special freedom that younger years and conventional circumstances could not provide. The very old may make some enemies or offend some acquaintances by their clarity and candor, but their close friends and family will not only understand, but may even admire and enjoy their company the more for it.

I find myself already on the way in this process. I no longer feel the need to be compliant to avoid making enemies. I am much more direct and outspoken about my ideas and opinions. I now have many acquaintances and fewer friends, but they are the ones who really care and count.

Women's Organizations:
Life After the Luncheons

The word "organization" meant different things to men than they did to women in my generation. To men, it meant their center of *business* activities; to women, it meant their center of *volunteer* activities. Men of my husband's age were involved in organizations — large, small, public, private, self- or other-operated — for work purposes. Women of my age, most of whom were not employed full-time outside the home, were involved in organizations for philanthropic or social purposes. Now that my generation has reached and passed retirement age, older men have become disconnected from their organizations, but older women have become, more than ever, connected with theirs.

I knew about the women's organizations in my community but, as a woman who *did* work full-time outside the home, I did not have time to participate in them, except for ritual financial contributions and vague appreciation for their good deeds. However, after I retired a few years ago and began to become involved myself, I began to learn many things I had not previously known about what these groups do, and especially about the women who do them.

What I already knew, but not as well, was how hard these organizations work on behalf of an array of causes, such as child abuse prevention, elder care, literacy improvement, environmental protection, and political awareness. They do this through a variety of volunteer services and fundraising activities, such as contests and cookouts, lunches and brunches, programs and projects, meetings and missions.

I also saw that these women not only work on behalf of *causes*, but on behalf of *each other*, as well. They provide transportation, or escort service, for members who need such help; they keep company with those who are ill or housebound; they monitor those who may need some assistance; they comfort those who suffer loss.

And finally, I found out that, at the same time as they are providing these services to others, many are also dealing with their own problems and pains, personal or physical, or both. They complain sometimes; they compare symptoms and circumstances with one another; but still, they continue to carry out their tasks, despite aching backs, aching feet and sometimes, aching hearts.

I remember reading an explanation that such older "organization women" are really using social or civic causes to resolve their unhappiness about no longer being needed, or to fill up their now "empty nests." My opinion is that this explanation is both untrue and unfair. It is true that older women may be finished with taking care of their young children, but they are far from finished with taking care of their families. Now, there may be frail aged parents who need care; or spouses with health problems; or grandchildren who need supervision; or adult children who need help with their lives. In fact, it has been estimated that women today spend as many years of their lives in family caretaking *after* their children are grown as they do *before*. Therefore, the organizational services women provide are not mere compensations or consolations for a "lost" mothering role, because that role actually continues; only now, the so-called "empty nest" is filled with different people and different needs than before. Their acts of giving and caring are in addition to, not in place of, family responsibilities.

Organization life provides opportunities for satisfaction, as well as service, because it enables women to help themselves, as well as to help others. Organization luncheons offer friendships, as well as food; and these friendships, in turn, offer opportunities for sharing together, as well as caring together.

For older men, on the other hand, there are fewer organizations (partly because there are fewer older men), and these are usually more task-oriented and geared to activities, rather than to services or to friendships. Sometimes, husbands try to "help out" in their wives' organizations, especially if they can be in the roles of fixers, arrangers, or managers. But they seem uncomfortable, or even embarrassed, when they do so, possibly because this places them in such an adjunct position in their wives' worlds.

Retirement life may be more difficult for men than for women, because women have so many more places and people to keep themselves actively occupied and satisfied. In many men's work lives, they had more job associates than personal friends; they had men there for shared tasks, but not for shared relationships. In my generation, making friends or needing friends — individually or within organizations — was accepted and expected for women, but not for men. For men, there was not only lack of time due to work, but also lack of facility and encouragement. There was also the assumption that their wives were sufficient as friends, as well as marital partners; in contrast, women have almost always had personal and social friends outside of, and in addition to, their marriages.

I wonder if it will be different for the next generation of young men, our adult sons. I notice that they seem to be more concerned about friendships and feelings than their fathers were (or than their fathers were permitted to be). I hope so, because otherwise, they too, like their fathers, will miss the special supports and satisfactions which women have enjoyed with each other, and from each other, over the years (and enjoy even more now, in their older years).

The Rocking Chair Traveler

More and more, I have been noticing special advertisements and arrangements for so-called "mature travelers," which really means senior citizens. Indeed, older people today are traveling more than they ever have, probably more than any other age group ever has, with the possible exception of the young back-packing crowd.

We seniors are traveling now because we finally have enough time, and many of us have finally saved enough, or are willing to spend enough, money. We want to travel to see certain places, some for the first time and some once again, before they change too much, or before we change too much. We know that, given the state of the world, the state of our finances, and the state of our health, there may be only a limited time left for us to do this. Therefore, as seniors, we not only feel that we can now afford to travel, but we cannot afford not to, because we cannot afford not to do whatever we are still capable of doing.

However, few of us are still bold enough or brave enough to travel far on our own, so we usually go on prearranged, or "senior," tours. That way, someone else has the responsibility for arranging itineraries, reserving hotels, planning transportation, and making decisions. It is interesting to note that the more "mature" a traveler we are, the more dependent we are willing to be. Another reason it is such a relief to travel with other seniors is that we can understand each other's desires and discomforts. We are not only heading for the same places on our trip, but we are also coming from the same places in our lives.

Indeed, the special circumstances of geriatric travel start even before we leave home, just with preparing and packing for our trip. Because we are so concerned about different climates in different countries, we pack light-, medium-, and heavy-weight clothing, plus insect repellent, sunscreen, windscreen, and lip balms. Because we are so concerned about sanitary facilities, we pack detergent, disinfectant, antibiotics, and toilet paper. Because we are so concerned about foreign food and drinking water, we bring our own cans and packages; but also, to be safe, bicarbonate of soda, antacids, laxatives, and diuretics. Finally, we add our history books, foreign language dictionaries, maps, itineraries, insurance pamphlets, reams of paper, and rolls of film. By the time we are finished with all this packing, we are over the allowed weight, over-prepared, and over-tired!

The next step is the journey itself, and arrival at the first destination. I don't sleep the night before I leave for a trip like this because of excitement; I don't sleep the first night I arrive because of the strange bed; and by the time I am able to sleep, it is time to move on to the next place. Even though flying times to faraway places are now shorter, my jet lag lasts longer. I find it hard to sleep during the long hours on the plane, but as soon as I arrive at our destination, I find it hard to stay awake. The first thing I try to locate at each new stop is not the nearest restaurant or museum, but the nearest doctor or pharmacy, just in case.

Even traveling on a senior tour can be demanding, but I don't dare to fall behind or miss a single sight or scene. For one thing, I really do not want to lose out on any opportunities; but also, I do not want to seem weaker than my fellow senior travelers. Deep down, I really appreciate the special arrangements — extended rest stops, reserved seats, and slower starts — but at the same time, I wish these aging accommodations were not so necessary, or so obvious.

During the trip itself, I enjoy seeing new and different places but, because there are so many of them, I may forget or confuse what I have seen. Meals in foreign restaurants are often served too late, and for too long and, for my needs, they often contain too much. Strange foods give me heartburn; strange beds give me backaches; and strange experiences give me palpitations. Foreign currencies are so complicated to compute that I can never figure fast enough to know whether I have paid enough, or too much, for something. Finally, there is the secret worry for all seniors about bathrooms — will they be accessible when needed; what will they be like; and will they work? As a group, we older travelers spend more sleepless nights imagining the trip in advance, and more sleepless nights afterward, recovering from them, than other travelers. When we were younger, we could better afford to travel physically, but not financially; and we were willing to endure more discomforts in order to enjoy more experiences. Now, the reverse is true.

So why do we seniors travel so much and enjoy it so much? Surprisingly, traveling in our older years is actually more satisfying than in our younger years. In addition to the pleasures of whatever we see or do, we enjoy the pleasure of knowing that we can still do it. As seniors, we are no longer very adventurous, if we ever were, so we do not go on mountain climbs or desert safaris or jungle treks; we do not need such exotic adventures, because just being able to travel is adventure enough!

We have all heard of the armchair traveler who never leaves home, and who views the world only vicariously. I would rather be a "rocking chair traveler," who goes as far and sees as much as possible, given the limitations of an older body with bunions, backaches and bowel problems. I would rather see some sights than none at all, and I would rather have to rest up after going somewhere than regret never having gone anywhere. It seems to me that the loss is not in not traveling far enough, but in not setting out on the journey at all.

Senior Volunteers

Being a volunteer means freely offering your personal and unpaid services to help others. We seniors provide more volunteer time than any other group in our population; in fact, it is estimated that approximately half of all volunteer hours are served by seniors. We support political candidates and causes; we visit the homebound; we tutor the illiterate; we collect money for the needy; we assist in understaffed classrooms; we console the bereaved and counsel the troubled; we write letters, make telephone calls and distribute petitions. We have more available time to offer, now that our job and child-rearing responsibilities are over; and we also have more skills to offer which we have learned over our lifetimes.

Being an older volunteer is a different experience for different seniors. Some, mainly women who did not work outside the home, have already been volunteers for much of their adult lives, in church, school or charitable organizations. But for those who *did* work outside the home, becoming a volunteer is a new choice and also a big change, and transferring from the work world to the volunteer world is not necessarily easy or automatic.

During our work lives, we often wished we could be free to choose what we wanted to do, when and how much we wanted to do, and to do it without compulsion or supervision. It is ironic that the new experience of volunteering, which provides these freedoms, sometimes seems as unnerving as it is liberating.

For one thing, we are unaccustomed to being unpaid after a work life in which our value was visibly measured in monetary terms. Our success and our status were demonstrated by the salary we received, the raises and bonuses we were awarded, and by the size of our office, equipment or staff. So now, as volunteers, how can our worth be recognized by others, and how can we ourselves know how valuable we are, if no money is involved?

Another difference is that volunteer activities are usually geared toward providing service, rather than toward producing tangible objects. So how can we judge our skill and success if there is no specific product or result?

Still another change is the flexibility of time and tasks after so many work years during which hierarchy and structure defined our roles and responsibilities. So, if each of us is now in charge of our own efforts, how can we tell whether we are meeting expectations, and how do we know how much is enough?

Finally, there is the need to prove ourselves all over again as volunteers after having already proven ourselves as workers. For those who were especially successful in their "other" lives at work, this may be both easier and harder to do. It may be easier because there are more skills and accomplishments to draw upon; it may also be harder now to do things another way, or another person's way. It may not be easy to serve on an equal basis with other volunteers who may be young enough to be our children or grandchildren, or who may not have our status, education or experience. On the other hand, it may also be more difficult to discover that we are not as up-to-date as we thought, and that newer methods and skills have superseded what we know, or thought we know.

Yet so many of us seniors are volunteers because it carries so many meanings and fulfills so many motivations. Our volunteer activities do more than just fill up our free hours; they provide structures and schedules that make certain days or hours of the week special. Also, volunteering is a use of our time that is socially useful and, therefore, socially acceptable. Our generation was raised under a work ethic that taught and required us to be occupied and productive, and that frowned on "mere" leisure. Volunteer activities are not only something to do, but something *worthwhile* to do, so that we are able to enjoy our leisure without guilt. In addition, it is a social activity because it is something we not only do *for* others, but also *with* others.

Another thing we enjoy about our volunteering is that it *is* voluntary, that *we* are the ones who make the decision to do it. At this time of our lives, when our choices seem more limited and we seem to be losing control over more things, it is comforting to enjoy, at least in this way, such a sense of choice and control.

We also feel that being a senior volunteer is an opportunity for us, as older people who have learned something during our lifetimes, to use what we have learned in new ways and for new purposes. When we grow older, most of us have a genuine desire to try to make some contribution or to be of some help and meaning to the world we live in, perhaps because we know that it is our last opportunity to do so. It is reassuring to realize that we may have retired from jobs, but not from life; that whatever wisdom we have earned and learned has not gone to waste.

A recent study of older volunteers found that they tend to be happier and healthier, as well as busier, than those who do not volunteer. It is not clear whether they become so as a result of being volunteers, or if they become volunteers because they are already this way. Perhaps it doesn't really matter which is cause and which is effect; what does matter is knowing that something that is so good for others is so good for us.

Smaller World, Wider World

Even though it sounds contradictory, our world grows both smaller and larger when we grow older. We become more detached about some things, but more involved in others. I suppose that's because we really live in two worlds, an outer one and an inner one. As our outer world of activities and responsibilities diminishes over time, our inner world of ideas and feelings expands.

Over the years, the boundaries of our physical world gradually contract. Retirement brings an end to involvement in jobs, businesses and careers and, therefore, an end to the need to travel regularly to places away from home. We also find that the less we leave home, the more difficult it becomes, and the less we are willing to do so. As a result, we don't get around as much, or go as far away as we used to; and so our own homes, our immediate neighborhoods, and our communities become the physical centers of our lives.

After we retire, we also often let go of our work friends, as well as our work, especially if these friends continue to work in the place from which we retired. For awhile, we maintain some contacts but, gradually, we begin to feel replaced and removed. The new people don't know or remember us, and we have less in common with the "old" ones we knew, because the work issues that still concern them are now only parts of our memories, not our current lives. Although we may follow some activities vicariously, or still attend some special events, our connections dwindle down eventually to receiving occasional notices or invitations and reading the company newsletters (especially the retirement announcements and the obituaries!).

We not only let go of our work associates, but also other acquaintances who live too far away. We become increasingly unwilling, or sometimes unable, to drive the distances, or expend the time and effort necessary, to maintain far-away friendships. First, our visits merely taper off in frequency; then they are replaced by telephone conversations; finally, even these dwindle, as loss of contact leads to loss of connection. Ultimately, the circle of our personal relationships consists of close-by, accessible neighbors and acquaintances, as our social world shrinks in number and geography.

In contrast, our "other" world actually grows larger because, now that we have finished attending to the cares of our outer lives, we can turn to our inner lives. When we were younger, we were occupied with earning a living and raising a family. We were caught up in daily, necessary, practical tasks and could not afford the luxury of introspection, curiosity or philosophy. (Perhaps, at that time, we also lacked the interest and perspective, as well.) But now we can allow ourselves these luxuries of greater freedom of meditation and imagination. At any rate, even though our bodies are less physically active, our minds are freer to wander and wonder.

We think seriously now about broad issues and events in the world, like poverty, peace, ecology, and justice, to name a few, that when we were younger, we probably regarded as too remote to be relevant. We not only think about these concerns, we take them seriously. We older people are sometimes accused of living in the past, but actually, we think a great deal about the future — even though we know it will not be our future. We know that it is too late to change our own world, or perhaps even our adult children's world, but it is our grandchildren's futures we worry about. In addition, even if we are unable to leave tangible gifts to our children and grandchildren, we want to give other kinds of gifts. We want to try to leave them a better world than the one left to us.

We find, at this time, that our thoughts not only journey further outward into the larger world, but also more deeply inward into ourselves. We begin to ask questions that we haven't, until now, usually, or consciously, asked. Why do we do what we do — or don't do? How do we really feel about other people, and how do we really want them to feel about us? What do we think we have accomplished so far, and how does that compare with what we wanted or expected? Now that we have finished growing up and have begun growing old, how do we sum up our lives?

This kind of inner thinking is sometimes described as being spiritual or philosophical. I am not certain I know what those terms mean, but I do know that we think about different things and in different ways than we ever did before. It is as if, having completed one part of our life's journey, we are now ready to embark on a different one. This time, it is a journey inward, in which we try to rediscover ourselves, perhaps for the last time in our lives, and we hope we will like what we find!

Sex and the Senior Citizen

The first statement I want to make on this subject is that just because we have snow in our hair or a chill in our bones, it doesn't mean that we have winter in our hearts! I've observed that there are two common stereotypes about senior citizens and sexuality; each is almost a mirror image of the other; and each one is wrong. The first stereotype perceives the older person as being virtually "de-sexed," — having no desires, feelings, or capacities for sex; the other is the pathetically, embarrassingly over-sexed older person — the "dirty old man," or foolish old flirt!

The first image assumes that because we've grown older, we should be beyond physical needs and feelings; as if suddenly we become both unloving and unlovable! Sometimes this also assumes such severe physical or mental decline that, even if we desired to be sexual, we would no longer be able, or know how to be. I'm sure we're all familiar with some of the jokes on this subject. For example, the elderly man who complained that he was still chasing women, but couldn't remember why! Or the elderly woman who explained that turning the lights down low was to save electricity, not to generate sparks!

On the other hand, when older people do enjoy their sexuality, this is considered unseemly. We've been so brainwashed that only youth is physically attractive that the image of aging bodies being sexual is ridiculous or repellent. As psychologist Judith Viorst commented, " ... the fires of passion are supposed to be either burned out or banked out." Either way, the message we're given is clear — young is sexy; old is not! It's interesting to note that, at the same time that younger people are demanding, and being granted, all kinds of sexual opportunities and liberties — in and out of, and before, marriage — we older people are being put out to pasture sexually. Why is it that it's the young who define what's natural or appropriate for us, not we ourselves, who have the actual experience?

However, sexual attitudes are changing a little, even for the older generation. Partly, this is due to some of the sexual revolution trickling down; mostly, this is due to changes in longevity and life expectations. We were brought up in a culture in which geriatric sexuality was not expected, or even imagined, let alone discussed. But we're younger and healthier than previous generations at our age and, although society may define us as old, our bodies are still only middle-aged, and know better.

I remember a few years ago telling my elderly father, who was then 90, that I had just read in Masters and Johnson's research on human sexuality that there was no reason why people could not be active until well into, or even beyond, their 80s! His response to this was a smile and a shrug, "Now she tells me!"

Now I would like to tell several things to those, especially members of the younger generation, who stereotype us as either neutered or nymphomaniac: First of all, the young did not invent, nor do they have any monopoly on, sexuality. We oldsters were enjoying satisfaction long before they even learned their first four-letter words! That's why I'm so amused when I hear these young people marvel about having "relationships" with their partners, as if this were some new discovery of theirs. Just what do they think my husband and I have been having for more than 50 years now, if not a relationship?

It's true that our sexual upbringing was different than theirs. We weren't trained for "recreational" sex, like some kind of contact sport; we weren't encouraged to seek adventure or variety, either of partners or positions. In fact, we were, and still are, dully monogamous and rather old-fashioned in the belief that sex and love are indivisible. Nowadays, as commentator Andy Rooney observed, love has been replaced by sex, as if there isn't room for both in relationships anymore.

Secondly, desire and pleasure don't inevitably wear out or vanish with age. Becoming older doesn't have to mean becoming colder, and being over 60 doesn't mean being over with sex. Our bodies as well as our minds are still being used and enjoyed. It's true there are changes, but these are only diminutions in frequency, not in pleasure. We outgrow the "first language" of the sex of our youth, which was concern for prowess and procreation. Now we move on to the "second language" of our older years, which communicates caring and intimacy.

Finally, if one has been lucky enough, as I have been, to still share life and love with the same partner, there is a special quality to that relationship. There is so much shared history and memory between us, there is so much intimate knowledge of one another, that we can exchange affection in a myriad of special ways. A smile, a touch, a caress can convey caring just as well as the most strenuous coupling. In fact, as we grow older together, our intimacy becomes even more precious because we realize that, in the last analysis, what we have is each other.

I suppose it all depends on whether one views the glass as being half-empty or half-full. Loving can last as long as living, and this is not a *problem* for us as we grow older, but a promise!

The Second Time Around

It has been said that when a spouse dies, "women mourn and men replace." This is not cynicism or criticism, but reality. It is not that men do not also grieve but, statistically, fewer men remain widowers than women, and even those do not remain so for very long. On the other hand, it is not that women do not also wish to have new relationships, but so many of them outlive men by so many years that widows tend to remarry less, or later.

It is estimated that, today, there are almost *four times* as many older women alone as men; and it has been noted that most of these women manage being alone better than men do. Perhaps this is because there are so many more of them to keep each other busy and to keep each other company. Perhaps it is because there are still so many more of the same things in their lives that they can still do, even though they are alone. Or perhaps it is simply because they know they must, since so many of them will remain alone for so long.

Yet, no matter how well they may manage, deep down, most widows and widowers do not really want to remain alone. Indeed, most of us, no matter what our age or condition, want to share our lives with someone else. The need is especially strong for those who have once known the experience of a life partner, because we know how important (even though sometimes imperfect) such a relationship can be.

It is not so much major events or exciting experiences we want to share with someone — because, if truth be told, there are not many of these in our older lives — but simply, our daily activities. We want to have someone to tell about what we did each day, who we saw or spoke to, what we bought, or even what we ate. We want someone to do things with, even if there is nothing specific or special we want to do. We want to have somebody nearby to worry about us, to help us if something goes wrong, or just to be there to know when, or if, something goes wrong. In fact, one of the greatest fears of living alone is that something may happen to us, and there will be nobody there to know about it.

Sharing with a friend, no matter how likable or sympathetic, is only a necessary, but inadequate, substitute for sharing with a mate. Even the most caring friends have other friends who occupy similar relationships with them; they cannot be specially and uniquely with us, or for us.

Yet older widows and widowers who do contemplate remarriage do so realistically, not romantically, because they know that there may be as many risks as rewards. First of all, remarriage means another major rearrangement of possessions, schedules, activities, and relationships. It means having to learn about, and learning to live with, a new person's wants and ways. It means having to take on some of another's lifestyle and letting go of some of your own, after so recently, or finally (with effort and pain and, usually, much of both) having learned to live by yourself.

Secondly, there may be discomfort about feelings of disloyalty or compromise, especially for widows for whom there is a more limited selection of partners available. There may be feelings of "settling for second best," as well as gratitude for a second chance. But widowers also, who compare a new present mate with the lost past one, can suffer nostalgia and disappointment. And for these "new" couples, there is also the discomfort of being measured against someone's memories, which means competing with ghosts; and in such a contest, reality can rarely win.

Finally, the most drastic deterrent of all, especially for those who have nursed a past spouse through a long and serious illness, is the fear of reliving that loss with still another spouse. As difficult as the experience was the first time, at least in a long-term marriage, it was preceded by many young years together. Because of such a long-shared past, the final caretaking is a painful, but not a resented, responsibility. But the second time, it is different, because there has been no long history together, and no time to accumulate one. The first time, we were young enough and well enough that worries about illness and death were far away until the very end. But the second time, given our age and health, we know from the very beginning that these possibilities may be too close for comfort.

As a result, there are some older men, and many older women, who remain alone as much by choice as by chance. There are some widows who disavow interest in remarriage, possibly because they know that circumstances and statistics are discouraging, at best. There are some who feel they can no longer make, or expect to receive, the same kind of total commitment they once did. There are some who fear that the second time will turn out to be an insufficient, or too-difficult, substitute for what they once had, or wished they had.

I suppose the fact is that remarriage for the elderly is bound to be *different*, not necessarily better or worse, because circumstances are different; and we ourselves are, by this time, not only older, but also different. The fact is that remarriage carries trade-offs. New relationships carry the cost of new risks; and new comforts, the cost of new complications. As many a widow would remark with a shrug and a sigh, she knew that either way, whether she remarried or remained alone, she would have some regrets.

A World of Difference and
A Different World

We can remember that when we were young, we had different ideas and attitudes than our parents, and that was expected because we belonged to different generations. We knew that, in time, our own children would also grow up to have different ideas and attitudes than ours. But I think that deep down, we expected that, because we were so well aware and so well prepared, the differences would be minimal or, at least, manageable. We expected that we would be better able to stand, and to understand, these differences than our own parents did with us.

However, things have not quite turned out that way! In fact, in retrospect, we now feel that we were less different from our parents than our adult children are from us. Our self-awareness has mainly served to make us more aware of the differences but hasn't given us a clue as to what to do about them. We know that different generations have different views of the world because they experience the world differently; but it seems that as we grow older, we find more in common with our parents' world than with our children's world.

One of the primary difficulties I've encountered is that the younger generation seems to speak a different language. Even though we may recognize the words, the meanings have changed, or certain words that once had neutral meanings have now taken on positive or negative connotations. A "relationship" seems now to be something adults experience before, or outside of marriage, rather than as part of it. For us, "open marriage" meant a public ceremony; now it means an arrangement without commitment. "Sharing" used to mean dividing time or toys, not baring our innermost secrets. It has been said that for our generation "closets were for clothes, not for coming out of; and grass was mowed, coke was a cold drink, and pot was something you cooked in..."

"Heroes" and "heroines" are now considered elitist terms so instead, we have "role models." "Compromise" denotes weakness, but "negotiation" is acceptable. "Agreement" is passive, but "consensus" is positive. We no longer have family or friends to assist us; we have "support systems." "Good" is whatever feels good, and "bad" is whatever does not win or work.

I realize that changes in language reflect changes in the times in which we live, but today's changes not only reflect *revisions* of our former lifestyles but almost complete *reversals*. Let me mention some examples of what I mean:

We were children of the Depression, so we expected and accepted limits in accomplishments, satisfactions, even possibilities. Today's generation seems to equate limitation with deprivation. We considered self-control and self-discipline as positive values; today, the goal is self-fulfillment, self-actualization or self-realization. We were trained to accommodate and compromise, and we prized courtesy over confrontation; now this courtesy is considered a "cop-out."

In our world, we accepted delayed gratification in our professional, personal, or public lives; and we believed, perhaps naively, that eventually, effort would earn reward. Today's generation is not as willing to wait, or do without, as we were. Our generation valued stability and security in our jobs, our living arrangements, our marriages; today's generation enjoys variety, excitement, experimentation.

We *feared* changes and choices; young people today seek them out.

We came from a generation which protected and preserved privacy. For us, privacy was not secrecy, denial or avoidance; it was simply that we placed certain boundaries around ourselves which limited what we were willing to expose to others and what we expected others to expose to us. We were not taught to "let it all hang out", especially negative feelings which we were supposed to hold within ourselves and handle by ourselves. On the other hand, current psychology encourages self-expression and self-exposure as confessional, cathartic and constructive.

The changes today are so complicated and so subjective that there are few certainties left, and what served us as moral clarities are now scorned as naivete. As a result, our generation is expected to accept behaviors today which only yesterday were considered unacceptable; sometimes we are asked not only to accept them but even to approve of them.

As the parents of today's generation, I realize that we helped to create or, at least make possible, many of these changes. We wanted our children to be able to enjoy liberties and luxuries that were not available to us. So when opportunities arose, we encouraged them; and we not only removed obstacles from their paths, but responsibilities, as well.

It is not that we regret making possible such a new world for our children, but that we no longer feel that it is our world. I really try to understand the new ways, but I continue to wonder whether too much of the past has been forsaken to make way for the future. I recognize that, as a member of the older generation, I feel more comfortable with what is more familiar; but I also know that being different is not always the same as being better.

On the one hand, my children's generation enjoys a wide range of opportunities and possibilities, material resources and technological advancements, role flexibilities and freedoms. On the other hand, they also suffer from separated and fragmented families, increased demands and pressures, complications and confusions. I am not really certain that I would want to live in their new world; I am not really certain whether I should congratulate them or console them —- probably both!

Women and "Girls" and "Men" and "Guys"

It seems to me that the way we refer to ourselves, and the way others refer to us, reveal a lot about the way we are seen and, sometimes, about the way we would like to be seen. For example, society calls us older people by many names — "senior citizens," "golden agers," "mature adults." These are all euphemisms that carefully avoid mention of the word "old." The message is clear. "Old" is not a desirable thing to be called, or a desirable thing to be.

We older people even give the same message ourselves when we refer to each other in ways and words that try to deny or conceal the reality of our own aging. For example, many women I know call themselves and each other, half self-mocking, but half wishful thinking, "the girls." When I am with them, they know that I object to the label so they try to humor me, although it's clear that they don't agree with or understand my protest. "It's only a word we use among ourselves," they tell me. "We're not serious; we don't mean anything by it, and we don't change anything by it. So what's the harm?"

Of course, we seventy-ish women know that we are not really "girls" anymore. So isn't it just a meaningless affectation or harmless bit of humor, as these friends insist? My feeling is that, just because it doesn't *change* anything doesn't mean that it doesn't *mean* anything. And the harm, I think, is not so much in pretending to be what we no longer are or can be, but in regretting what we no longer are or can be. So, in one sense, it is meaningless, because we know that it doesn't make us any different; but in another sense, it means that this is the way we *wish* we could be different.

I understand that, for us older women living in a society that prizes youth and beauty, we miss our lost girlhood; and in a society that has become so different and so difficult, we miss our lost innocence. But wishing that we and the world were the way we used to be won't bring back either one. So it is not an exercise in nostalgia, but in futility.

For myself, I don't see myself as a "girl," or want others to see me as one, and I certainly don't want to be one again. I wouldn't want to go back to the past and have to do it all over again the way I was then, and I wouldn't want to have to do it all over again the way the world is now. In fact, there's a certain relief for me in knowing that I don't have to do either one.

Chronologically, my girlhood is now more than half a century behind me. Since then, I have been a young woman, a mature woman, a middle-aged woman, and I am now an aging woman.

Physiologically, my body certainly knows that I am no longer a girl. My waistline is thick, and my blood is thin; my pressure is up, and my resistance is down! I suppose it would be nice to be dark-haired, slim-hipped and firm-fleshed again. But I know that it's more important to me now that my mind stay firm than my figure, and that my views and vision are growing broader, along with the rest of me!

Psychologically, I am certainly not a girl, either. It has taken me years, and not always easy ones, to become a grown-up woman, and I'm relieved that I've gotten so far and, relatively, so well. The lessons I have learned over all those years did not come without cost, so I would not want to have to undo them or go through them again. I suppose it would be nice to feel youthfully footloose or frivolous again, but not if it means letting go of any of the commitments I have formed or the understandings I have finally found. I can remember the girl I used to be and still recognize some of her in the woman I am now. I certainly don't want to lose her or forget her—I just don't want to *be* her!

I have noticed that older men rarely refer to themselves as "the boys;" in fact they find that label foolish and even embarrassing. When men are together, I hear them refer to each other as "the guys," not "the boys." Perhaps, in their memories, "boys" are powerless creatures, while "guys" are manly, strong and independent. I understand that, for these older men living in a society that is run by, and mostly respects, the powerful young, they miss their lost strength, not just physically but psychologically. So it is not that our men are any less prone to nostalgia about lost youthfulness than we women are, but that they miss something different. For them, it is not looks or innocence that they long for, but the lost sense of power.

Perhaps you may be thinking that I am making too much of a few mere words, supposedly said just in jest. Of course, these labels cannot really change who or what we are now. We remain not only men and women, but aging men and women, not the "girls" or "guys" we may call ourselves. But when we continue to mourn the qualities we have lost over the years, we may overlook the qualities we have gained over those years. And when we keep wishing or pretending—though not intentionally or seriously— to be the "girls" or "guys" we once used to be, we may not be able to appreciate the grown-up men and women we have finally become.

Part III

All in the Family

Long-Term Marriage, Long-Term Mate

My generation may be the first, and also the only one, to celebrate long-term marriages of 40, 50, or even 60 years! We are the first because our parents and grandparents didn't live long enough to reach such advanced anniversaries with their spouses; and we are probably the only one because our adult children are marrying later, or divorcing and remarrying, or not marrying at all; so, although they are living longer, it is not with the same spouses.

As more and more of my generation live longer and stay married longer, "golden" anniversaries become milestones, but no longer miracles. As long-term mates, we have now lived more than twice as many years together as we have apart. Consequently, the "70- year itch," rather than the "seven-year itch," has become a stepping stone! Are our lifetime marriages merely year after year of the same thing, or do they change over time? And if they change, do we also change with them? If so, which is cause, and which is consequence?

As I look back at my own 50 years of marriage, I am aware that many things have happened to my husband and me during that time. Our children have grown up and gone, leaving us without an active parenting role. (In fact, our years of the "empty nest" last longer than those of our full nest!) We have both retired from work, leaving us without active professional roles; so now we have more time at home and more time together, but less time ahead of us than behind us. We have seen our parents age and die; and we have become the older generation. Now, when we look at each other, we see the lines and signs of aging, and we begin to glimpse our own mortality. It's ironic that we spend so much of our married lifetimes trying to change each other, when it's really these shared experiences that reshape us both.

What keeps us together through all these changing years is as much reality as romance, i.e., an inseparable combination of fears, familiarities, and facts.

For one thing, we worry about being left alone in our later years, without anyone caring about, or sharing with us. It's not only concern about having someone to care for us, but also having someone for us to care *about*. For instance, I've noticed that I worry much more about my husband's health than I ever did before. Maybe that has to do with the new reality of health problems, or with ensuring my own possible need for caretaking, or with an unconscious fear of widowhood. All I know is that, with the children gone, my attention has shifted from being child-centered to being spouse-centered.

We also stay together out of sheer familiarity. Over the years, we have forged a fundamental alliance which is stronger than the little daily damages we may do to it. We understand each other's unspoken languages, and can hear the messages even before they have been expressed. We know each other so well that we know which "buttons" to push, and also which ones to avoid, ignore, or respect. We know each other's needs, and have even become accustomed to our incompatibilities. In fact, even our differences seem to diminish over the decades. When both of us and our marriage were younger, we had disagreements about money, children, power, and goals. Now that these issues have either receded from our lives, or in one way or another been resolved, we find that what we have left are more bonds than barriers.

That's not to say that there still aren't frustrations and flare-ups. In fact, one woman described her feelings about her long-time husband as, "Divorce never, but murder frequently!" These are the feelings that become counter-balanced by what psychologist-poet Judith Viorst called " ... a long history — connections that help render us complete — ties that hold us and heal us."

In addition, there are the facts of life and marriage as we learned them when we were growing up. Nobody promised us that we'd always be happy, so we didn't expect magic and sunshine forever. Our children's generation has sophisticated marriages, and also sophisticated divorces; we weren't so sophisticated; we just stayed married. For us, compromise is not a cop-out, and accommodation is an acceptable art. We know that what we have is what we have to work with; and we don't waste time thinking that somewhere else, or someone else, will be better.

So with time and effort, and the help of humor, we learn to accept each other as we are. I know that my husband will not really change anymore and, for that matter, neither will I. In fact, by accepting him, I enable him to accept me. In other words, we're finally able to reconcile the marriage we once thought we wanted with the one we really have, and not feel disappointed or regretful about the result.

As I look back at our journey so far, I can still recall some of the twists and turns, and even near-accidents, we encountered on the way. But in spite of them, or maybe even because of them, I would like my husband to know that I look forward to more miles and more milestones together.

Relating to Elderly Parents: the Aging and the Aged

Today, families are more vertical than horizontal; they contain more graying generations than growing ones. As a result, more and more of us "senior citizens" still have our own elderly parents, who are now in their 80s, or even 90s. This means that we are the *older* generation, but not the *oldest*; at the same time that we are parents and grandparents, we are still children to our elderly parents.

We are the first generation of seniors to find ourselves in this situation of being both *parents* longer and *children* longer. It is a role for which we were not really prepared, and for which there are no precedents or guidelines. On the contrary, we find ourselves having to make our way through uncharted emotional minefields. The next generation will probably have it easier, because we will have established models for them. Also, for better or for worse, we have freed our own children from the same overwhelming sense of obligation to the parents for whom we accept responsibility.

The expression "sandwich generation" has been commonly used to describe our new situation, but I think it is actually different and more complicated than that. The "sandwich generation" was initially perceived to be caught between the competing needs of aging parents and growing young children. But there has now been a major age shift toward the later part of the lifeline. Our elderly parents, in their 80s or 90s, are truly frail and old; we ourselves, in our 60s or 70s, are no longer the middle generation, but have begun our own aging process; our adult children, not us, are now the ones at the crest of their lives and on the threshold of their middle age; and it is our grandchildren who are now the young generation. So, instead of being the middle filling, we have become one of the outer layers in a *double-decker* sandwich!

As our now-grown children move further away from us through geography or independence, our now- elderly parents move closer to us through the reverse process. If truth be told, we would rather spend more time with our children and, especially, with our grandchildren; but as columnist Ellen Goodman commented, this is the stage of life when we have to worry more about parental care than childcare. We know it is a blessing to have our parents live so long, but sometimes, the blessing feels like a burden. In our younger years, we were child-centered; now, in our later years, we are parent-centered. It makes us wonder if we will ever have a chance to be self-centered!

There is a common myth that elderly parents in our society are abandoned or neglected by their families. The truth of the matter is that many adult children, unless geographically distant, see their elderly parents at least weekly, and provide most of the services and care they need. In fact, adult children are now providing more care, and more *difficult* care, for parents over a longer period of time than was ever done in the past. My own elderly parents died only a few years ago, and my husband's 90-year-old mother is still living. I join him in spending time with her, helping with housekeeping tasks, managing finances, monitoring her care. Deep down, we want to do right by our parents, even though we may not always know how. Indeed, it is precisely because the myth of family uninvolvement is so untrue that we adult children find ourselves facing so many pressures and feeling so much pain.

As we watch the changes in our aged parents from the once-powerful figures of our past, we remember, and miss, the parents we knew — our "real parents." There is a part of us, no matter how old we are, that still feels like a child and wants to have strong, capable parents to help us and protect us. Instead, we now often find ourselves having to take care of those who used to be our caregivers. In a sense, we are the ones who feel abandoned by the parents we used to know, rather than the other way around.

Watching what happens to them also forces us to "preview" our own aging process; their mortality reminds us of our own. We know that our aging is different from our parents' — it is less advanced and less irreplaceable, and we also have more knowledge and more resources — but even so, we recognize in them what may be preordained for us.

In addition to these feelings of loss and fear, there are also feelings of helplessness and guilt. We see our elderly parents in the last stages of their lives, and there is little we can do. We cannot turn back the clock; we cannot make them young and whole again; we cannot "save" them. Our helplessness, in turn, often makes us feel guilty, no matter what we do or don't do. Somehow, we feel that we are not doing, or cannot do, as much for them as they once did for us. It has been said that it is "the debt we can never repay."

Dealing with our elderly parents can also bring out "ghosts" from our family closet, which holds all of our old childhood resentments, angers, and disappointments about the way we were raised. Some psychologists suggest that we try to resolve this "unfinished business" by airing our grievances while our elderly parents are still alive. I don't agree; in fact, I don't even think there is any point in doing so. We need to extricate ourselves from what psychologist Eda LeShan called "the mirage of perfection" and recognize that, just as we cannot be perfect children to our parents, they could not be perfect parents to us. In a sense, the "unfinished business" actually is finished; because our parents are now no longer the "enemy," and we are no longer the "victims." Neither of us is who or what we once were; and the past must finally be filed away and forgiven.

Another notion we hear often, and with which I also disagree, is that of role-reversal with elderly parents. It's true that we are at the stage where our parents probably need us more than we need them; where we now have to perform certain tasks which they formerly performed for themselves, or even for us. But even though the responsibilities have shifted, our roles have merely changed, not reversed. No matter how old we, or they are, they are still our parents; and as long as they are alive, we remain their children.

One of the things that strikes me is that I am now older than my parents were when I thought they were old! That makes me aware of my own possible future dependency needs, and makes me wonder how my own adult children will deal with them when that time comes. Therefore, how we now treat our elderly parents in the sunset of their lives becomes a role model for our children for the way they should someday treat us.

We know that the relationship with our parents is one of the few in our lives that is irrevocable. Sure, we can have an ex-spouse, or an ex-friend, or ex-boss; but we cannot have an ex-parent! Indeed, our connections with our parents often surprise us with their power and their persistence. Rabbi Max Vorspan once commented that we may spend a lifetime trying to cut the umbilical cord, but somehow, in the end, we find ourselves becoming like our parents anyway. With a sense of shock, I see myself doing things and hear myself saying things that my parents used to do and say. I look in the mirror, and I recognize the resemblance; I look down, and I see my mother's hands; I look around, and I become aware of the inheritance I have never escaped. Perhaps, in the long run, becoming like our parents is our final way of accepting them, and of accepting the ultimate loss of them that we will one day have to face.

The "Golden Oldie"

My husband and I have just celebrated our golden wedding anniversary, but celebrate seems such a simple word for such a complex experience. Although we certainly did celebrate, it was more than just enjoyment. It was a complicated combination of reliving the past, recognizing the present, and reflecting on the future. It also felt like an incredible contradiction of reality and unreality.

On the one hand, I cannot believe we have been married so long but, at the same time, it is hard to remember any time we were *not* married! I cannot believe that so much time has passed, because it seems to have passed so quickly; yet I know it must have, because so much has happened to the world and to us during those years.

We have lived through World War 11, the Cold War, the explosion of technology and population and longevity, and decades of change and turbulence. We have lived through changes in our own lives, as well as in the world; our children have been born and have grown up; now our grandchildren are growing up; our parents have died; there have been places traveled and visited, homes built and lived in; there have been professional accomplishments; and there have been illnesses overcome and losses experienced.

But if so many years have passed and so much has changed, why do I still feel so much the same, as well as different? Why do I feel that we are not that much unlike the two young people who met and married half a century ago? When I look at my husband, he doesn't seem so changed from what he used to be; and so I can believe, or pretend, that I am not, either. But when I look at pictures from the past, I can see the marks of all the time that has passed; and then sometimes when I look at him, I see his father's face. When he looks at those same pictures, does he also see the differences in me? And when he looks at my face, does he see my mother, as well as me? It comes as a shock that we have changed so much when it feels like it's hardly been any time atall.

I remember when my parents were at this age and stage of life, they seemed to be so much older; although their lives were not yet finished, they seemed to be. They seemed to have reached a kind of plateau or state of completion that appeared to be beyond change. But my husband and I do not feel that we are finished growing, just because we have finished growing *up*. We both feel there are still things we want to do, and places we still want to go, and changes we still want to make. Did our parents feel the same way, only we just didn't see it? Or are we truly a *younger* older generation than they were?

In the past, people were not married as long primarily because they didn't *live* as long—or, if they did, they were usually in such frail health or poor condition by that time, that it was scarcely a time for celebration. But now, in the retirement community in which we live, 50th anniversaries are not such a rare or remarkable occurrence. In fact, the long-married couples who attract attention there are the ones celebrating their *60th* anniversaries; 50 years of marriage is no longer such a big deal.

But for us, whose lives and marriage it is, it *is* a big deal. We have now lived together more than twice as long as we have lived apart, and our marriage spans almost three quarters of our lifetime. We started out half a century ago as a young couple together; then, for a few decades, we were parents together; and now, for the last half of our long marriage, we have once again become a couple together, only an aging one. I cannot believe that we are as old as we are and that we have been married as long as we have; but when I wonder where all those years went, I realize that they went into what we have now become.

It has been said of long-term marriages that people don't really stay married to the same person for all those years, because each of us changes over all those years. That makes our anniversary even more of an accomplishment— as well as an astonishment—because we have had to learn to live with changes in ourselves—as well as in each other and in our circumstances.

I don't know what I expected it to be like when we started out 50 years ago and, of course, I cannot know how many more years there will still be for us together; I can only know how far we have come together. I also know that, especially during these latest years, we have grown not just accustomed to each other, but indispensable to each other. We know that our children and grandchildren, as they should and must, follow their own separate paths; so, although they are always part of our lives, they are not *partners* in our lives. That means it is the two of us, my husband and I, who are the ones who live together and last together.

This golden anniversary, regardless of how unreal it may, at times, seem, has now become the *most* real thing in our lives.

Becoming Grandparents, or "Having an Enemy in Common"

Being a grandparent may be the only relationship we have that's all gift and no grief! Parents, children, spouses, and siblings all seem to bring us as much pressure as pleasure. But somehow, grandchildren feel like a biological gift, requiring little or no effort on our part (that is, if we overlook all the work involved in having raised their parents!). It's the one family relationship that is without responsibility, and so it is without disappointment or guilt. Somehow, we don't worry so much, or feel so threatened, about being good grandparents as we did about being good parents.

In fact, it's been said that one reason we grandparents and our grandchildren get along so well together is that we "have an enemy in common!" We are natural allies against a single target, the middle generation. For both of us, the generation in between represents a source of stress, either past or present. As we have learned from experience, being a grandparent is easy; it's being a parent that's difficult!

Our grandchildren are not only gifts in themselves, but they also *bring* special gifts. First of all, they give us a second chance to be the kind of parent figures that we didn't have the time, patience, skill, or resources to be with our own children when they were small. Through our grandchildren, we relive our early experiences, but this time, we do everything right!

I think this is especially true for the grandfathers who did very little child-rearing with their own children. I've watched my own husband baby-sit, feed, and diaper our granddaughter as he never could, or wanted to do, with our own small children. His lack of parental involvement then was partly because of his work attachments, and partly because it was not then considered a male role. So now, he is eager for this last chance to parent. As one child recently described it, "A grandfather is really a man/grandmother!"

Our grandchildren also give us uncritical acceptance which we receive from no one else. They enjoy our company, our stories, and our attention without minding, or even noticing, our shortcomings or our changes — unlike our children, who *always* do! To them, we look and behave exactly as grandparents are supposed to — not aged, but ageless.

Our grandchildren give us a sense of continuity and immortality, although we probably don't use, or even think in, those kinds of dramatic terms. But they do make us feel linked to the future. We feel this in the special pleasure of recognizing familiar traits and resemblances, as if parts of us are continuing in them. In fact, my husband and I sometimes even compete with each other about which one of us our granddaughter looks or acts more like, as if the greater the resemblance, the greater the chance of immortality.

Finally, our grandchildren just bring us a gift of sheer joy. They reawaken in us a feeling of fun, and they are the only ones who make us smile totally, both inside and outside.

But it's not just a one-way connection, because we give precious gifts to our grandchildren, as well. We give them unconditional love, with no strings attached. They don't have to seek it, or earn it, or perform for it in any way. It is a love that they are simply given for being our grandchildren. We have the wonderful freedom to be different from their parents — not strict or serious, not nervous or nagging. We can afford to spoil them by breaking the rules, because we don't *make* the rules. Grandparents are for saying "yes," not "no," to their grandchildren!

Also, we give them our gift of unmeasured time. Because we're retired now, we don't have work commitments to engage us; because we're usually visitors in each other's homes, we don't have domestic responsibilities to distract us. Our time and attention is undivided, and all theirs. One little boy remarked about grandprents that, "everybody should have them because they're the only grown-ups who always have time for you."

We also give our grandchildren a sense of history; just as they provide us with the future, we provide them with the past. We tell them special stories we have saved and savored about how things used to be. There's sheer pleasure in sharing these reminiscences with them, and hope that some will be remembered, along with some memories of us.

Finally, we can give our grandchildren a cushion of constancy in a world, and among people — possibly including even their own parents — that seems to be always changing. I don't mean just financial help, or shelter, or advice (though we give all these, and they are all important), but the special certainty that we will remain the same for them, even though other things may alter, or falter.

We should keep in mind that every time a *grandchild* is born, a *grandparent* is born, too! Even though, on the outside, we grandparents of today may not look or behave like those from past generations, on the inside, we still share the same powerful and instinctive feelings. It's not only that these are our children's children; they are also our grandchildren, and we have our own direct and unique connection with them.

In a recent survey of modern families, family researchers Kornhaber & Woodward found that the bond between grandparents and grandchildren is the second most important human connection, following that between parent and child. Perhaps it's because we are not the primary relationship in each other's lives that we're free to enjoy one another in ways we never could, and never will, with either our own parents or our own children.

Living Longer, Grandparenting Longer

Grandparents today are not the same as those of the past. First of all, there are more of us living now — approximately 50 million — than at any previously reported time. Secondly, we are living longer. As a result, we continue to be grandparents for more years and for a larger part of our lives than any preceding generation. In fact, we now spend approximately one-third of our lives in the grandparent role! Finally, we also remain younger longer and become older later. That means we now become grandparents not at a younger *age*, but at a younger *stage* of our lives. Overall, we are younger-looking, healthier-feeling and more diversely active than any previous generation of grandparents.

Because we are younger-looking, our grandchildren may view us as only slightly older and slightly heavier versions of their parents' adult generation, rather than as "old" people. In fact, it has been observed that today, we look like "regular" people, not like grandparents! This can present to our grandchildren a positive role model of aging, rather than one that is frail or frightening.

Because we are healthier-*feeling*, we are more physically fit not only to do more things for ourselves, but with our grandchildren, as well. We are able to go with them to theaters, restaurants, and museums; we go hiking, biking, and back-packing; we go on visits, voyages, and adventures together. We are able to be active participants in our grandchildren's experiences, rather then merely passive providers of "love and cookies."

Because we are more diversely active, we are retired, but not retiring. We are busy traveling, volunteering, taking courses, pursuing creative hobbies, even starting second careers. Our schedules may be too full to permit unlimited or instant time availability, but they are filled with interests and ideas that we can share with our grandchildren.

Another result of living longer is that our family lines are now lengthened on either side of us. We may still have our own elderly parents, as well as our young grandchildren. Four-generation families are becoming increasingly commonplace, which means that, to our grandchildren, we are an older generation, but not the oldest. We now find ourselves in between as "younger oldsters," and, in this middle position, we are the ones who form the linkage that binds the generations. We are the ones who keep our grandchildren connected with the family continuum.

At the same time that we are growing older during these added years, our grandchildren are also growing up. In the past, it used to be that grandchildren were usually only little ones; today, we have the possibility of living long enough to enjoy adolescent, or even young-adult, grandchildren. This means that we may experience different stages, as well as different ages, as grandparents.

In the beginning, when our grandchildren are small, we serve primarily as playmates, baby-sitters and story-tellers. We experience sheer joy in being with them, watching them, and "spoiling" them, without responsibility or guilt.

As our grandchildren grow into adolescence, we can serve other functions. We then become interpreters, negotiators, "escape hatches" for them, and for their parents, as well. Indeed, in many ways, it is easier for grandparents and growing grandchildren to relate to each other than to the generation in between, which represents for both of them "the enemy in common!"

Later, as our grandchildren move into young adulthood and approach the threshold of their own independent family lives, we can serve still another purpose. We can be a source of family history and collected memories. We can also share with them whatever wisdom we have won over the years and offer them advice for their own futures. Perhaps because, by this time, we have lived such a long time, our suggestions are now more welcomed and carry more weight.

Finally, if we are lucky enough to live long enough, we may be able to enjoy the ultimate experience of becoming great-grandparents. This means that we will be able to relive the entire life cycle one more time as we experience the role-shifting of generations, as our grandchildren become young parents and our adult children become grandparents themselves. Psychologist Judith Viorst called grandparenting "parenting once removed." Therefore, great-grandparenting, which is removed even one more step from primary caretaking, can provide even more pleasures and possibilities, and even less pressures. Someone commented that it is this final stage that really adds the "great" to grandparenting!

As we now have more years and more ways to be grandparents, we find that each new role is a new opportunity, without any loss of the old. As we watch our grandchildren grow from childhood to young adulthood, we can offer them different gifts to meet their changing needs — pride, support, history, wisdom. And as we move from our own middle age to final old age, we also receive from our grandchildren different gifts to meet our changing needs — joy, caring, respect, continuity.

These added years in themselves are not the gift; what they make possible is. Being able to be grandparents in new, different and longer-lasting ways is a gift that keeps giving, not only for us as grandparents, but for all the generations in our family.

Being "Blended" Grandparents Without Getting "All Stirred Up"

Becoming a grandparent used to be simple and predictable because there was only one way to do so — we married, had children who grew up and, in time, themselves married and had children of their own, who were our grandchildren. Today it is different; it is easier to *become* a grandparent, but more complicated to *be* one.

Today, through our children's marriages to spouses who have been married before, we can become instant grandparents to grandchildren to whom we are not biologically related in any way. The so-called "blended" family — consisting of previously married parents, plus an assortment of his, hers and their children — is the most rapidly growing marital arrangement today. This is not only a new kind of family form, but "blended" grandparents are a new kind of relationship within that form, since each of these families contains at least four different sets of differently related grandparents.

In all the extensive current literature about "blended" families, there is virtually no attention paid to grandparents. Yet we know that although our adult children may end their marriages, we do not end our role as grandparents; there may be a divorce, but we never get divorced from our grandchildren!

As a result, as "blended" grandparents, we must now relate not only to our adult children and their new spouses but, in some cases, the former spouses of each, the children of each of their separate marriages, and any children in common of their current marriage. Indeed, these arrangements may be so complicated that the term "blended" (suggesting a smooth mixture) is probably a misnomer; "stirred up" is more like it!

But no matter what name we use, the "blended" family is so new and so different from the past, although so increasingly common, that we have no clear guidelines, or even the correct vocabulary, for it. What should we call our new non-biologically related grandchildren, and what should they call us? How should we refer to our new son- or daughter-in-law's former spouse, who is the other parent of these grandchildren? How should we — or *should* we — distinguish between the different sets of other grandparents — those related to our own grandchildren, and those related to our step-grandchildren? We need new words to define all these new relationships.

Trying to be a grandparent to these different kinds of grandchildren is made even more complicated by the variety of living and visiting arrangements. These usually depend on whether our adult child is a daughter or a son, and on which one is the custodial parent of the children from the former marriage. One result may be a continuing close relationship with our biological grandchildren, but only limited and occasional contact with the other grandchildren. However, another result may be that, realistically and regretfully, we have more access to our step-grandchildren than to our "real" grandchildren, and that we become relegated to being only periodic, or "special-occasion," grandparents.

Whatever the arrangements, relating to these different sets of grandchildren is like practicing a delicate juggling act, trying to balance treatment that is similar with feelings that are different. For example, there are certain concrete or practical issues we have to deal with. On birthdays, holidays, or special occasions, should we give the same gifts, or gifts of the same value, to both sets of grandchildren? If they have different interests or hobbies, should we make certain to spend the same amount of time on each, and offer the same amount of praise? Should we try to make everything absolutely equal, as if there were no differences between the two sets of grandchildren, even though we all know there really are?

We do not, and cannot, have the same feelings toward our step-grandchildren who are not connected to us through biology or history as we do toward our "real" grandchildren. Nor can they duplicate with us the same relationship they have with their own biological grand-parents. Yet we want to "adopt" these new grandchildren — emotionally, if not actually or legally — because, no matter how tenuous or complicated the relationship, we are now related to each other; we are now members of the same family.

We also *feel* for these children, because we know that they have experienced the same kinds of family problems and partings that our own grandchildren have. We may not be able to love them in the same way, but we do want to help them in the same way. We also hope that, by helping our step-grandchildren, we may also be helping our adult child's new marriage, as well. Finally, we know that our own biological grandchildren may be experiencing uncer-tainties with *their* new step-grandparents. So we hope that the way we behave in our new role in the blended family can be an example for other grandparents in "adopting" our grandchildren.

This is certainly not the kind of simple, gratifying grandparenting we expected and looked forward to; nor is it the kind of grandparenting we saw our own parents enjoy with our children. Indeed, we may be the first and only generation of grandparents faced with this special situation. In the past, families usually remained stable and, therefore, so did the grandparents' role. In the future, divorce and remarriage may be so commonplace that *all* generations will be accustomed to changing parents and grandparents.

But meanwhile, our own generation of grandparents is still struggling to sort out old relationships and establish new ones. Although being a blended grandparent today is so different and so difficult, it may be even more important than ever, because grandchildren are more than ever in need of safe and stable adults in their lives. The important thing for us, as blended grandparents — with *all* of our grandchildren, both biological and non-biological — is that we never allow our role as grandparents to end, for any of them.

The Last Grandchild

My latest, and most likely last, grandchild was recently born. Not being the first one, the sense of novelty is less, but the sense of this child being a miracle is as strong as ever. In fact, having experienced grandparenting several times already actually enhances my anticipation and pleasure, because now I know what I have to look forward to.

There has been an interval of some time since our other grandchildren were born; they are now beginning to grow up, so this latest one seems like a special and unexpected gift. Being a late grandchild brings both advantages and disadvantages. It means that he will be able to live to see a later future than any of the others; but it also means that I will probably not live to see as much of his life as I will of the others. I do not expect to be able to see him grow up, and I regret that I will not be part of many things in his life, but this makes whatever I *am* part of even more precious.

My daughters live far away, so when each of my grandchildren was born, I packed my "grandmother bag" and went to be with them to help; and I did so this time, also. While I was there, I supervised the other children and the housekeeping and the meals. I tried to provide my daughter with as much rest and support (and as little advice) as possible. I also tried to provide her with some private time with her husband and older children, which gave me the opportunity to enjoy my own private time with the new baby without taking him away from parents or siblings, or being restricted about not spoiling him.

I also enjoyed quiet moments alone with my daughter when the baby was asleep and the others had gone to work or school. I enjoyed preparing all her favorite foods, which I was convinced she had not eaten, or eaten enough of, in years; and I fussed over her in ways I had not been able to for a long time.

I remember that I was cautioned by everyone not to try to do too much; but of course, I did. Yet the truth is that those who worried about it being too hard for me did not realize how easy it was in a different way. And those who made a point of how helpful it was for my daughter did not realize what a gift it also was for me.

Being there offered me a special and last opportunity not just to do helpful, "grandmothering" things, but also to do "mothering" things, which my daughter's adult competence did not usually invite or permit. For this brief moment, we could both go back in time; I could once again be her nurturing mother, and she could be my accepting child.

It seems to me that our need to parent our children far outlasts their need to be parented by us. Our children become so adult so soon. We are proud of how capable and independent they are; yet we also remember — with fondness, nostalgia, and some regret — how they once looked up to us and needed us, and how we were able to take care of them. Now that they are parents to their own children, they have become the ones who are protective and nurturing.

My daughter did not resist or resent my parenting the way she had as an adolescent or young girl; nor did she complain that I was being overprotective or infantilizing. Perhaps my caretaking was acceptable because we both knew it was so time-limited, and also because it was so related to an immediate and specific reason. It was not some vague, devious attempt on my part to exercise control; nor was it some serious subservience or regression on her part. I hope that, during our time together, while I was enjoying the chance to be her parent again, she too, deep down, was enjoying the chance to be my child.

While I was there, mothering and grandmothering, another scene also played before my eyes. In this remembered vision, it was my daughter who was the brand-new child, and it was I who was the new young mother. When each of my now-adult daughters was born, my own mother came to stay for a while, to do "grandmother things" for them and for me. I remember how I appreciated the physical relief her help provided; but even more, I remember how I appreciated being fussed over, worried about, and taken care of myself, especially since I knew that I was now responsible for taking care of my own child. I wonder if my mother thought and hoped then that someday her daughter would be doing the same for her own child, just as I now hope for my daughter and grandchild.

When I came back from my grandmothering visit, I was physically exhausted, and I jokingly complained that I was too old for such "new grandmother things." But deep down, I knew that I was not really objecting, but actually rejoicing, and I would not have missed this wonderful, wearying experience for anything. I knew that this special, and probably final, opportunity to give something to my child and my grandchild was a gift to me, as well as to them, and also an opportunity to connect with both my remembered past and my expected future.

Author Marilyn French recently wrote a novel about the relationships among mothers, daughters, and grandchildren. At the end of the story, the heroine observes, with almost a sense of wonder, that the thing that really endures is "... this continuity, this line ... these people in my life who can't be changed, exchanged, or substituted for." I know exactly what she meant.

Sibling Rivalry Revisited

When we think of siblings, we usually picture young children playing with each other, or competing with each other for some desired prize or praise. But we are siblings all of our lives — not just when we are young — even after we grow up, or grow old, or grow apart. Indeed, our sibling relationships may be the longest-lasting ones of our entire lives; longer than our oldest friendships, or our marriages, or our relationships with our children; even longer than our relationships with our parents, because we usually outlive them.

We may choose our friends, our spouses, and even our enemies; but with our siblings, we do not get the chance to choose the ones we want. In fact, they may not be the kinds of persons we would have voluntarily selected, and each of us may wish that the other were different. But though we may not be built-in friends, we *are* built-in family; and although we did not choose each other, we cannot really lose each other.

Our relationships with our siblings are shaped by our actual shared experiences, as well as by our perceptions and memories of those experiences. If you were the oldest sibling, you probably enjoyed the power over younger sisters or brothers, but resented the responsibility and sense of replacement newer children bring to those already born. If you were the youngest sibling, you probably enjoyed being pampered and privileged, but resented the enforced subservience to older sisters or brothers. If you were the middle child, you probably resented both the younger and the older ones.

We remember the sibling who was always held up as the ideal for us to emulate, but whom we could never equal; we remember the sibling who seemed to enjoy such an easy relationship with our parents, compared with our own, thorny one; we remember the sibling with whom we competed for attention or approval, and who invariably won both! It's as if we have kept a running account of ancient grudges — at least in our minds — even though, by now, we have probably forgotten the actual events.

Even after we and our siblings grow up and our parents grow old, we tend to remain in our same roles with one another. It is distance, not time, that makes a major difference in our relationships; siblings who are geographically close tend to remain personally close, while long-distance siblings eventually become emotionally, as well as physically, separated.

When our elderly parents die, we suffer not only the personal loss of them, but also the loss of the primary unifying force that bound us together as siblings. While our parents are alive, we all maintain the "family front;" but when they die, there is no longer anyone to serve as "kin-keeper." We are then faced with new choices about our relationships with our siblings. Do we merely allow them to remain as before; or do we try to reinforce and rebuild them; or simply let them run down and, eventually, run out?

I have known siblings who have not seen each other in years, except at compulsory family occasions; or who have not spoken to each other, except for minimal cursory exchanges. I have known siblings who are still trying to settle old scores through new adult achievements, or through spouses' successes, or even through another generation of rivalry among their children. I have even known siblings who have severed all contact with each other.

These may seem extreme situations, but they occur often enough to form what one psychologist called "a family tragedy."

Yet, both because of and in spite of all that has gone on between us, our relationships with our siblings persist and, in fact, actually become more important, not less, as we grow older. During childhood, our connection is physically close because we live together, but it is filled with rivalries and resentments. During adulthood, we become separated, as we are each preoccupied with developing our own families, careers, and autonomies in different ways and in different places. But during our later years, we often feel the need to return and reconnect with the earlier parts of our lives, which includes our siblings.

Especially after the death of our parents, we realize that our siblings are the only ones left from our family of origin. They are unique because they are the only other human beings who shared our old home, our history, and our heredity. They are the only ones who are part of our vanished past, and who can still remember the children we once were. Only my sister can remember the fat, funny little girl I used to be, and how I almost broke her brand-new doll carriage by trying to climb into it. Only my sister is also a child to my parents, and can share the same mixture of love and loss, remembrance and regret that I feel about them.

As siblings, we may not have picked each other, or preferred each other, or always pleased each other; but that doesn't really matter anymore. At this stage of our lives, when there is so much less left, our connections with one another are more important than our conflicts. We cannot rewrite or relive our family history; neither can we return to earlier days, to either heal or avenge old wounds. What we *can* do is let go of the image we have of either a hated rival or a hoped-for ideal relative (the former probably never really existed, except in our imaginations; and the latter probably never could have existed, in reality). Now, we and our siblings can forgive each other for whatever we could not be for each other, and finally accept what we really are — imperfect but important people in one another's lives.

Our Siblings and Ourselves,
Growing Older

At this stage of our lives, we come to realize that our siblings are the people in our lives whom we have known the longest, and who have known us the longest. Our parents, who knew us first and best, are now gone. Our spouses – even of 40, 50, or more years – who are now the ones who know us best, were not part of the earliest part of our lives. Even our oldest and closest friends were not with us at our beginnings. It is only our siblings who are still left from our earliest lives; and even though they may no longer be our closest or nearest relationships, they are the longest-lasting relationships in our lives.

Although we may live far apart from our siblings now, both in life space and lifestyle, and although they may not play an important or active role in our present lives, they are an important part of our past lives. In fact, the longest, and probably the strongest, memories we have of them are from our childhood, when we were an ever-present part of each other's lives.

The forces that shaped our relationships with our siblings – from gender, to position in the family, to personality, or even to our physical resemblances – formed patterns that still seem to persist in our minds today, as well as in our memories. Indeed, it is those childhood patterns that we seem to remember most vividly now, in spite of the miles that have separated us, or the years that have aged us, or the experiences that have changed us.

Even though we know that our siblings have changed, just as we know we have, it still comes as a surprise to us to see the aging persons they have now become. I think this is especially true if our siblings do not live nearby, or if we see them only occasionally, rather than regularly. The gaps between our contacts are filled with mental pictures of the way we all once were, so we grow accustomed to thinking of them as we remember the way they were, and the way we were with *them*. In a sense, seeing the way our siblings have now aged and changed is like looking in a mirror, because we know it also reflects how *we* have aged and changed.

So it seems strange to see that our once-rambunctious "kid" brothers, or once-pesky little sisters, whom we used to protect or patronize, have now become such graying and slowing older men and women. And it seems strange to see that our "big" sisters or "big" brothers, who once taught us or bossed us around, have now become the ones who need our help, or cause us concern.

My own sister is several years older than I am, and the difference in our ages was usually too much for her to be my friend, but enough for her to be my role model. As a child and teenager, I could watch the steps she went through in her life experiences — dating, marrying, becoming a new mother — and get a preview of what would happen to me.

When we were both young adults, the age difference seemed to grow less as we began to have more in common and went through similar experiences as wives and mothers. In fact, for awhile, when we became mature adults — mothers-in-law and grandmothers — the age difference almost vanished as we shared the experiences of our parents' dying and our own children becoming young adults.

But now that we are both older women — one just beginning, and the other ending, her 70s — the age difference has once again begun to widen, as I cross the threshold of old age, and I see that she is already there! I notice now, with concern and surprise, that her step is no longer as steady, her body is no longer as straight, and her voice is no longer as strong. In fact, she has become, once again, the role model, to show me what my future will be.

In our generation, illness and fragility are no strangers, and eventually, neither is death. But the loss of a sibling – no matter how old, no matter how close or distant, and no matter how often death has occurred to others we knew – is like losing a part of our own lives, a part of ourselves. Who will be left now to remember us as the children we once were? Who will be able to remember our parents the way we do? Who will there be who lived in the same home, tasted the same foods, heard the same stories, or were taught the same lessons as we were? We still have our own memories, of course, but without someone who shared them to confirm and relive them with us, they become mere ghosts.

When we lose a sibling, we lose not just that particular person or relationship, but perhaps our last remaining link with our past. Our siblings are special people in our lives. Sometimes they are support, and sometimes they are stresses, or even strangers, to us. But it does not really matter because, regardless of whether we live like each other — or even like each other — we have intimately and irrevocably shared and shaped each other's past, and have been, at least partially, shaped by them. So, when our siblings age and fail, our own lives are changed, too. And when our siblings die, we know there is no one else — no matter how close to us they may be — who can bring back the particular part of our lives we have lost with that sibling.

The Disneyland Child, or
"Absence Makes the Heart Grow Fonder"

The expression "Disneyland parent" refers to the non-custodial parent who entertains his child on holidays or vacations, and enjoys the pleasures of parenting without ongoing responsibility. For adult siblings and their elderly parents, there also seems to be a comparable "Disneyland child." This is the long-distance adult child who visits periodically and enjoys filial feelings without ongoing contact or caretaking. In fact, this is occurring more and more, as our elderly parents live longer and our adult children live further away. As a result, there may be two sets of children — the close-by and the far-away — who have different relationships with their parents, and with each other, as well.

The far-away child is the one who sends expensive gifts and sentimental birthday cards, and makes long-distance telephone calls; the close-by child is the one who gives immediate time, effort, and care. Each one often resents and envies the other; the close-by child envies the other's freedom from family responsibility, and the far-away child envies the other's family closeness and connection.

Whenever the long-distance child comes for a periodic or "special-occasion" visit, this event becomes the overriding topic of conversation for the elderly parents beforehand, and the overriding memory for them afterward. The parents also worry about the expense of these long-distance visits, and the time and exhaustion involved, but not so much that they may discourage or diminish the visits.

During these times, the long-distance child is usually

regarded as a special guest of honor who is invited, but not involved. Few demands, expectations, or pressures are applied to this "guest," as if by virtue of visiting, filial obligations are fulfilled. On the other hand, the near-by child feels suddenly secondary and excluded. The more frequent, but less dramatic, expenditures of time and money elicit little comment, let alone concern. Caretaking activities are taken for granted or ignored, but are still expected to continue.

Also during these infrequent visits, there may be brief times of private, and sometimes uncomfortable, conversations between the siblings. The visiting sibling often makes concerned observations about their parents' changing and aging, as if presenting new descriptions or discoveries. These comments come from concern, but invariably state what is already known to the sibling who has stayed close-by. The effect, though not the intent, of these comments is to make the other sibling feel incompetent or insensitive. Along with the observations, the visiting sibling often makes suggestions about possible things the other sibling could do or try to help their parents, like hiring house-keeping help, arranging for attendant care, spending more time or money, finding a different doctor, or pursuing different medical treatment. These suggestions may be well-intended and delicately phrased, but they usually are ideas already considered, or even unsuccessfully attempted. Although presented as suggestions, they often feel like instructions, or even indictments!

The visits of the long-distance sibling are time-limited, as well as periodic; therefore, they finally end. Afterward, the parents usually feel regret, but the local child usually experiences mixed feelings. Ultimately, they all resume their regular routine of contact, caretaking, and coping — until the occasion of the next visit.

While my own parents were still living, I was the close-by, caretaking child, while my long-distance sibling, who lived on the East Coast, came to visit twice a year. I remember the discomforts and resentments that I felt. I envied my sister's exemption from the emergencies, panic telephone calls, canceled social appointments, and uneasy vacations that I experienced with our parents. I resented the fuss that was made over the presents that she sent, or her presence when she visited. My sister and I still live geographically apart; since our parents' death, we have continued our semi-annual bicoastal contacts. During these past few years, as I have thought about our differing relationships with our parents, I have come to some unexpected, but comforting, conclusions.

First of all, I have come to realize that surprisingly, in some ways, mine was actually the easier sibling role. My sister was relieved of responsibility, but also of the intimacy of my relationship with our parents; she was less inconvenienced, but also less needed; she was left out of my parents' problems, but she was also, largely, left out of their lives. On the other hand, I was able to share not only my parents' problems, but also their pleasures; I was able to be an important part of whatever they remembered, not just what they regretted. It was as if my role in my parents' lives was real, while my sister's was remote, or even irrelevant.

Secondly, I finally understand that each of us was only trying to express her own caring in whatever way was available. Since I was close-by, I could be involved in a steady sequence of small helping activities. But since my sister was present only periodically and briefly, her participation could only be limited. That is probably why she felt such a need to be dramatic and directive when she was here — to make up for all the times and ways she knew she could *not* be here. It was as if she needed to make her presence felt in order to make up for her absence. It has occured to me more recently that we are *both* our parents' children, and we were only coming from different directions geographically, not emotionally.

Our Other Siblings

Most of us have other siblings, in addition to the ones who were born into our families. They are the brothers and sisters of the spouses we marry, as well as of the spouses our brothers and sisters marry.

By this time, we have known these other siblings for as long as we, and they, have been married, sometimes as much as 40, 50, or more years. I have known my sister's husband for more than half a century, and my husband's two sisters for almost that long. I remember my brother-in-law when he was a shy, slim, dark-haired young man, and my two sisters-in-law when they were round-faced adolescents. We have known each other for virtually all of our adult lives; we are of a comparable age and stage of life, as well as duration of marriage. We have a great deal in common, but also a great deal that is not.

We know our in-law siblings longer than we know many of our friends, but we still may not actually be friends. We know them almost as long as many of our relatives, but we may still not really feel like members of the same family. It is a long-term, but often ambiguous relationship; one that endures and impacts our lives in *some* ways but, in *other* ways, is never as close, or as crucial, as with our "real" siblings.

For one thing, these in-laws were already adults — or almost adults — when we first met, so we do not share any common childhood history. We have no memories or experiences of each other before the time we first met as adolescents or young adults. We are relatives created by law, not by life.

For another thing, these in-laws entered our lives involuntarily; they were chosen by, or brought into our lives by, our siblings, because of *their* needs or connections, not *ours*. As a result, we became acquainted with them, and usually related to them, indirectly, through those others with whom they had their primary relationship — our own spouses or siblings. That means that our relationship with them was probably dependent on, or even determined by, those other intermediate relationships, like how well our spouse got along with his/her siblings, or how well we got along with our own siblings. Our early reactions were almost always defined by other people — not by ourselves or each other, at least initially; and also, initially, they probably seemed more like complications than complements to our lives.

Our relationships with sibling in-laws may proceed along many different kinds of paths. We may find ourselves feeling competitive with each other about our, or their, spouses' or children's accomplishments or acquisitions. Or we may find ourselves competing for the attention or approval of the relatives we now share through our, or their, marriages. On the other hand, we may find ourselves becoming allies with each other against other relatives in our shared in-law families. Or we may sometimes find that, due to similar interests, personalities, or geographic proximity, we actually feel closer to an in-law sibling than to a "real" one. If that happens, we feel guilty for liking the in-law more, and for wishing that our "real" sibling was more like our other one.

Regardless of which form our relationship follows, we know that it is different than with our biological siblings. We have fewer expectations of our in-law siblings, and therefore, fewer disappointments. We hope mainly for a relationship that is civil and sensible, not necessarily positive but, at least, not negative; so, if it turns out to be better than that, it is a gift. We know that we will never care about them as much as we do about our biological siblings, but we will also never be as angry at them, or as hurt by them.

If they encounter problems in their lives, we feel distressed for them, but not despairing. If they enjoy successes in their lives, we feel pleased for them, but not triumphant. It is as if we experience each other in a kind of second-hand way, through an emotional filter; we are like siblings one step removed from each other.

But, like so many of the relationships in our lives, this one also changes in our older years. By this time — and it may have taken all of us a long time — we have been able to work through to our own perceptions and relationships with each other, which may be apart from, and different from, those we had initially.

Also, during all these years, we have shared family celebrations — birthdays, anniversaries, graduations — and family sufferings — illness, failures, deaths. We have been part of each other's milestone events, and have been available to help each other when needed, perhaps not out of the same kind of innate emotional investment as with our natural siblings, but still out of real concern and connection.

We have become accustomed to each other during all these years of sharing our families' growing up, and now we are sharing our own growing older. We have come to the stage where we appreciate one another, not just for being what we are, but for still being at all! In a world that is changing so much, and in which we are losing so many we know, those who remain with us become an even more important part of our stability and our support system.

PART III, ALL IN THE FAMILY

So it turns out that we are neither strangers nor siblings, neither friends nor family, but a special combination and variation of all of these. We would probably be considered supporting players, not leading roles, in the cast of characters of each other's stories. But even though we may not be *necessary* to one another's lives, now, in our later years, it is certainly nice to *have* one another in our lives!

Family In-Laws and Outlaws

There is no single word, or even simple expression, in our vocabulary that defines the relationship we have with the parents of our married children's spouses. For want of any other word, we usually use the term "in-laws," which is really neither clear nor accurate. Yet it is a relationship which, if we are lucky enough to have our children's marriages last long enough, will last for a large part of our lives. The fact is that when our adult children marry, a whole new set of relatives is introduced, not just into their lives, but into our lives, as well.

This relationship with in-laws is not only unusual, but sometimes even seems unnatural. For one thing, it is entirely involuntary; we have no choice in selecting them, but are nevertheless compelled to accept them. We may know nothing about them, except that they are the parents of our children's chosen partners; yet we are expected to admit them, even welcome them, instantaneously into the intimacy of our family.

Not only do these new relatives feel like strangers, but sometimes like intruders, as well. We used to be the only parents in our children's lives; now there is another set of parents who have suddenly appeared to claim the same role we have been occupying. They expect similar attention and affection or, at least, respect; they expect to have their own special needs and preferences noticed; they expect to be included and consulted no less than we are in family decisions and discussions.

These parents were not there all along, during all those years and decades (not always easy ones) when we were struggling with our children's growing up. Now, when our children have finally become reasonable adults, we feel we deserve, and would like to enjoy, the fruits of our new and calm relationship with them. Instead, we find ourselves forced to share these hard-won benefits with strangers who have done nothing to develop or deserve them. The fact that they may see us in the same situation with their children is not something we usually consider.

In addition to being strangers and intruders, these in-laws may also be rivals to be reckoned with. They compete with us for access and influence with our children; they expect equal time, attention and treatment. The competition becomes even more keen over grandchildren, over such questions as what to name the child — for which relative or side of the family; or over whom the child resembles, which child-rearing practices to follow, or whose baby gifts are more useful, desirable or expensive; and later, about which grandparents the children see more often, or like better.

Yet, over time, we change, and we notice changes in our relationship with our in-laws, as well. Together, we watch our children move into beginning middle-age, and our grandchildren grow up to become the young adults we remember our children being when we first met one another.

Now we share feelings of relief and appreciation that our children's marriage has lasted — by this time, 15, 20, or more years — especially given the current divorce rate in their generation. This is a feat for which both sets of parents probably take some private credit, either for their intervention, or their non-intervention, as the case may be. We are grateful that our children and our grandchildren have been spared the pain and problems of broken homes and broken hearts. We are also grateful for ourselves, that we have been spared the strain of experiencing our children's pain.

By now, enough years have passed that we have come to know each other well enough to work out an arrangement, which may vary from being an unarmed truce to actual friendship. We have now each probably carved out certain special family areas, or roles, which are mutually accepted or respected. There are certain occasions which are the established domain of one or the other set of parents; there are certain activities or entitlements which are specifically assigned; there are certain tastes and requirements which are acknowledged. We have learned to recognize and avoid each other's taboos, and not to threaten each other's vulnerabilities. Perhaps we are now able to accept each other's special areas of influence because we have also established our own. We also find that the more we are able to accommodate, the less we need to compete. We probably realize that though we do not have to be friends, neither do we have to be enemies.

Over the years, we develop a shared history with these people. We accumulate common family memories, like births and birthdays, graduations and promotions, disappointments, and even deaths. We have survived problems in our children's and grandchildren's lives, and in our own lives. We may even sometimes have offered each other assistance or support at such times, so that we now feel like survivors together.

Finally, we not only share a common past, but we now also share a common future. We notice that we are all growing perceptibly grayer and slower. We know that they can see in us the same signs of aging that we see in them, as we share the winding down of our lives. We can sense what each other is experiencing in ways we could not when we were all younger. At the beginning of our relationship, we were more different and less realistic; now, over the years, we find that we have more in common than in conflict. Perhaps we may never have become friends, but we are certainly no longer strangers. Perhaps what has happened to us is that we have finally become each other's family.

The Generation Gaps

I used to think that the expression "generation gap" pertained only to the differences between young people, usually adolescents, and their parents. Now I know that there is more than one generation gap — or perhaps it is still the same one, only continued over time. I didn't realize that this gap between children and parents would persist no matter how old the children and the parents were. I didn't expect that I would someday find myself connected with, and conflicted with, two other generations — my grown children and my aged parents — and that I would become caught in a generation gap on both sides.

Why is it that the people I can relate to most easily are those to whom I am the least related — my friends, who are members of my own generation? On the other hand, with the people who are so important in my life — my parents and children — why do I have the most complicated connection? It is as if we are too close because of our kinship, but also too separate because of our generations; so the ties that bind us also burden us.

With our parents and children, we have a relationship that is vertical, not horizontal; that means that we can never be at the same place in our lives at the same time. Now that I am at the same rung on the generational ladder as I remember my parents being, my grown grandchildren are where I used to be. That means now that I can finally understand my elderly parents' feelings, I realize that my adult children still cannot really understand mine.

I remember that when I was with my elderly parents, I often found myself involuntarily reverting to the old childhood roles with them of feeling obedient and ingratiating, or rebellious and resentful. But when I am with my adult children, I also seem to regress and involuntarily return to the past role of giving instructions and directions, or of expecting dependence and obedience from them. It is as if, with both of these generations, I have never totally forgotten, or freed myself from, our long-ago roles with one another.

And it becomes even more complicated, because it is sometimes difficult to separate what used to happen between my parents and me from what still happens between my children and me. These two sets of relationships seem to overlap, or parallel, or even imitate each other. When I speak to my children, I sometimes find myself, with a combination of shock and memory, saying things my parents used to say to me, in almost the same tone, and with the same words! Or sometimes, I realize that I am doing the opposite — trying to recreate my own childhood in reverse with my children, only not as it actually was, but as I fantasized it could be.

I think there are two main reasons for these stresses between our generations: One is that we know these ties are so irrevocable. They are not optional or voluntary, or dependent on any contingencies, allowances, or preferences. We do not choose our parents or our children, and are not chosen by them, and we know we can neither reshape them, nor replace them. We can never become ex-parents to our children, or ex-children to our parents, no matter what our ages, or wishes, or circumstances may be.

The second reason is that we know these ties are so powerful. Whatever we say to one another and whatever we do to one another, intentionally or otherwise, has the power to hurt us or to heal us. Whatever mistakes we make, of omission or commission, or merely of oversight, can affect each other's feelings, or even change each other's lives. That is why we often find ourselves having to watch our words or monitor our actions, because we feel so vulnerable to being hurt by them, and so fearful of hurting them. Even in the extreme situations in which parents and children have become estranged from each other, the very absence of that relationship is, in itself, powerful in their lives.

An interesting thing I have noticed, however, is that if the generations are removed even one step from each other, as we are with our grandchildren, the stress and the gap grow less. With our grandchildren, we can still be close, but not too close for comfort; and somehow, we can enjoy one another without the burden we feel with either our parents or our children. It may come as a surprise for us, and even a source of envy, to realize that our parents probably enjoyed a similarly smooth relationship with their grandchildren, who are our children, which we could not with either of them.

It seems ironic that this should be so, that when we are more removed from our primary connections, whether with parents or with children, the more relaxed the relationship can be. Whereas, with our parents and children, we know that both the bond between us and the gap between us will always remain, in spite of all other relationships in our lives, in spite of all other changes in our lives, and in spite of all the time that passes in our lives.

Part IV

Parents and Children:
Beloved Enemies

Adult Children: Friends, Foes or Family?

The phrase "adult children" seems to me to be a contradiction in terms. If they're adults, they're responsible and detached; if they're children, they're dependent and emotionally involved. Why is it I feel both ways? Chronologically, I know that my grown daughters are adults, but emotionally, they remain my children. For me, being the parent of grown children has been filled with contradictions. I kept wanting them to grow up but, when they did, I felt they had grown up too fast, and had moved too far away. I wanted my grown children to be independent and effective; but I also wanted them to ask my advice and accept my values. In other words, I wanted to have both an "empty nest" and a "full nest" at the same time!

I may have thought that being a parent of small children was difficult, but at least *then*, I knew what was expected; now, I'm not sure. I have the feeling that we're all struggling to redefine our roles with each other. Am I supposed to be simply an older, wiser friend? Or do I continue to be a monitoring, nurturing parent figure? Am I expected to be parental occasionally and selectively, at designated times of need? If so, who defines the need?

I've noticed that when my daughters come for a visit, we all tend to regress to the roles we carried when they were children. In my mind's eye, and in my home, they are not adults, but children, who still require my protection and direction. It's as if giving advice is the last vestige of active parenting I have left. Sooner or later, there's some disagreement, which is resolved with regrets, and also relief that we're just visiting each other, not living together.

On the other hand, when I visit in *their* homes, I am compelled to recognize them as separate adults; but that isn't comfortable, either. I see their different lifestyles, and I feel torn between what psychiatrist Erik Erikson called "concern that prompts us to advise, and respect that prompts us to refrain." I used to think that when my children grew up and left home, I wouldn't have to worry about their problems anymore. But when I'm there, and see things happening in their lives, the old feelings of protection and responsibility come flooding back; only now that they're grown, I can't do anything about it. Who said that being the parent of an adult means being a childless parent!

Longevity is an issue now, also. My adult children are in their 30s and 40s, approaching the threshold of middle age; and I'm older now than my parents were when I thought *they* were old! In the past, when children were middle-aged, their parents were either dependently old, or dead. Now, they are our 40- and 50-year-old "children," who worry about their own growing children, and yet, still have functioning parents who worry about them. The whole process is so prolonged that we now spend more years as parents of adults than as parents of young children.

Now, there are also grandchildren, who can be *another* complication. I remember years ago having revenge fantasies about what would happen to my children when they had children of their own! Sometimes, I had hopes that they would "come back" in a different way when they became parents themselves. But I sometimes feel as if my adult children envy the indulgent relationship my husband and I have with their children, which they don't have with either one of us. Do they resent the time, attention, and affection we lavish on their children, compared to what they received — or what they *perceive* they received — when they were small? Do they feel that we love our grandchildren more than we love them? It's not true, of course; it's just that being a grandparent is a spectator sport, while being a parent is more like hand-to-hand combat!

My daughters have been long gone from my child-free home, but I still feel their absence, and their presence. I keep wondering why it is that with other young adults, who are not my *own* children, I can enjoy relationships that are natural and honest? Why is it, instead, that with my own grown children, there are so many things not spoken or understood, so much "unfinished business"?

I think much of this comes from our expectations of each other, which are probably unfulfillable. I know now that I will never be the "perfect" parent my children wanted me to be, or that I, myself, wanted to be for them. I also know now that they can never be the ideal children who will fulfill all my dreams and needs. So when will we forgive each other for not living up to our fantasies about parents and children?

Erma Bombeck called this difficult, complicated relationship "... family ties that bind... and gag!" At any rate, my grown children and I seem to have the enduring capacity to dismay and delight, to exasperate and exalt each other. Whether we like it or not, we are locked together for life! As I grow older, I wish there were a way to let them know that, when all is said and done, I would not have it any other way, and I would not have *them* any other way. Judith Viorst said it so well in her poem entitled, "Before I Go":

"Before I go ...
I'd like those I love to know that they are the ones,
If I could do it again, I'd do it with."

When is it Our Turn?

It seems ironic to me that, years ago, when *we* were the younger generation, it was advantageous to be the older one; and now that we are the older generation, it is the younger one that has the advantages. It is not that any formal or official rules have changed over the years, but expectations have changed. As a result, for us, no matter on which side of the generational relationship we have been — as young adults relating to older parents, or as older parents relating to adult children —we have been the ones who are expected to *give* and to give *in*. As a matter of fact, until a few years ago, while my elderly parents were still living, I found myself having to accommodate both other generations at the same time!

Let me mention some examples of what I mean. First of all, when my husband and I were married, we had the kind of wedding our parents wanted — formal and filled with family. But a generation later, when our daughters were married, they had the kinds of weddings *they* wanted — filled with friends, folk music, and free-style vows.

I remember that, as young adults, when my parents or my husband's parents came to visit, we strained ourselves for their convenience and comfort. We not only tried to do things that would bring them pleasure and win their approval, but we also tried to protect them from any worry about possible problems in our own lives. Now, when we visit our children, we feel that almost the reverse is happening. We are the ones, not they, who are careful not to cause any inconvenience. We are the ones who try not to present any problems, who try to protect them from any worry about us, even if there might be a realistic cause for worry.

I remember that I used to adjust my personal schedule to make certain I would have enough time for my parents; I still need to adjust my personal schedule, but now it is to be available to my children when *they* are free. When I was young, I deferred to my parents' authority; when I became a parent myself, I deferred to my children's needs. In the past, I worried about being a "good girl," who would meet my parents' expectations; now I worry about being a "good mother," who can meet my children's expectations.

Another difference is the new ideal of so-called honesty and openness between the generations. I used to refrain from disagreeing with my parents, and certainly from criticizing them. Sometimes, this meant having to deny or suppress some of my own feelings, but I preferred that to confrontation. Mine was a generation that found it easier to feel guilty than angry. On the other hand, my own children have no hesitation about being "totally honest" with me, which usually consists of pointing out my mistakes, failures, or shortcomings.

Another consequence of this new candor is that our adult children no longer feel the need to pretend about anything in their lives, or to protect us from any problems in their lives. Certainly, we want our children to be able to be honest with us and not to be limited to the kind of restricted relationship we sometimes experienced with our own parents. And certainly, our children's candor is not intended to cause us any pain or make any demands on us, yet both of those result. I also know that all this honesty is supposed to make for more mature and meaningful relationships between us. But sometimes, I find myself wishing that, once in awhile, even if only for awhile, I could exchange the "true confessions" for some blissful ignorance!

The irony is that we have come to realize that we parents are responsible for some of these very changes that we have now come to regret. Because we were raised to feel such a sense of obligation to parents and obedience to persons in authority, we did the opposite with our own children; we liberated them from the constraints that bound us. We assured them that they did not owe us any debt, because we were their parents, so they were free to pursue their own purposes. We minimized our expectations of them, so that they could maximize their own attention to themselves.

Now, over time, we are beginning to wonder whether we were wise. How were we to know that we might someday ourselves want some of the same kind of solicitude that our parents received from us? How were we to know that we might come to regret some of the permissions we gave our children, which they took; or some of the lessons we taught them, which they learned? We did not expect to suffer second thoughts about being such emancipated and emancipating parents.

However, the process is not complete because, lately, I have noticed that time is bringing new transformations, as the generations are once more shifting. Our adult children, who are now entering middle- age, are seeing their own children take their former place as the younger generation. They are moving into a relationship with their children comparable to what ours was so recently with them. I wonder if they, too, will experience some of the mixed feelings that we did. I wonder if they will begin to appreciate, or at least better understand, what it was like for us. I wonder, as they move toward becoming the older generation, whether they begin to see themselves in us. And I also wonder, as each of us moves into a different "turn" in the relationship between the generations, whether we will now move closer to each other.

When the Shoe is on the Other Foot

We wanted our children to grow up to be caring, considerate adults, especially with us. We also wanted them to understand and appreciate at least some of what we did for them as their parents, when they were unable to do things for themselves. We expected this to evolve within the context of our customary parent-child relationship, the only difference being that we would all be older by that time. What we wanted was what psychoanalyst Erik Erikson has called a "new, different love of one's parents," which may finally occur during adulthood.

Indeed, I have recently noticed that our two adult daughters display increasing considerateness toward my husband and me; but this is also mingled with increasing concern, even though we have not knowingly invited or implied any such need. They are more concerned about our health now; even innocuous aches and pains elicit worried phone calls, well-meant remedies, or urgent requests to see doctors. They increasingly ask tactful questions about our medical care and coverage. When they are with us, they don't let us lift heavy packages, or walk long distances, or drive too fast or too far. They worry about our working too hard or doing too much; and they earnestly assure us that we have earned the privileges of pleasure and ease in our later years.

I know that it is intended to be comforting that my children care enough to want to help; but it is also discomforting that they think I may need that help! As a result, I don't know whether to be amused or annoyed, flattered or frightened, by this new solicitude. I am amused because our children seem to think that they now know what is good for us (as we once did for them!) But I also feel annoyed because it seems critical of my current competence. I am flattered because it demonstrates their caring and appreciation; but I am also frightened because it suggests that they may be seeing some incapacities in us of which we are not aware. Parents are never pleased to become dependent, or even to feel dependent, upon their children, no matter how old any of them may be. I still see myself as strong and capable of parenting; does this solicitude mean that they do not see me this way? My grown children may be ready to relieve me of responsibilities, but am I?

This concern often becomes more evident after some major event in our lives, like retirement; change of home or location; the death of our elderly parents; or any injury or illness we may have incurred. Perhaps these major life changes make us seem more vulnerable to our children because they represent significant departures from what we used to be.

But sometimes, there seems to be no visible event in our lives; so the change in our children's attitudes toward us may be due to something unseen in their lives. Perhaps, as they cross the threshold into middle age and begin to feel the first faint twinges of their own aging, they recognize in us their own futures. Whatever the impetus, the transition is usually gradual, moving from small, simple, occasional gestures to worry and protectiveness. But no matter how gradual, it still comes as much as alarm as comfort to us when our children make it known that they are now the ones to offer, and not to receive, assistance.

This may be especially difficult for us if we ourselves have only recently completed the caretaking of our own elderly parents. We are reminded of our recent responsibilities for aged parents whom we remember as feeble and failing, inept and in need. We worry, then, about whether we are beginning to appear to our grown children as our parents appeared to us. We wonder when it was that we began to become our parents?

In the past, older generations rarely experienced this kind of transition, because they usually did not live long enough to see the emergence of middle-aged children. But my generation is witnessing the aging of our own children, who now begin to look the way we so recently remember ourselves; and it is the grandchildren who look like, and are, the young generation.

Our adult children are not only moving into middle age, but also into the central family role we have, until now, occupied as primary caretakers, decision-makers, and power sources. We watch this transition with both pride and pain. We feel proud of the active, capable, responsible adults that our children have become. But we also feel the pain of realizing that, as they move onto center stage, we now move to the sidelines. I wonder if they have any mixed feelings, as we do, as they assume the roles and responsibilities we are now abdicating or losing? Along with the sense of satisfaction they must feel for themselves, do they also feel any sense of loss for us, for the powerful parent figures we once were?

We wanted our children to be free of responsibility for us when we grew older, but not free of relationship with us. We wanted affection and appreciation from them, not assistance. We wanted them to care about us, not to have to take care of us.

A friend recently observed, with a mixture of pain and puzzlement, "When I could no longer mother my daughter, I had to mother my mother. I suppose someday, my daughter will also have to mother me!" I know that my generation liked it better when we were still able to be adult children to our own parents, and adult parents to our own children.

Wearing Out the Welcome Mat

When our children grow up, we expect that they will live apart from us, but I am not sure we expected it would be so far apart. Sometimes, for reasons of education, careers, spouses, or personal preference, they move many miles, or even, as in the case of my own two daughters, a continent away. Sometimes, we parents are the ones who relocate after retirement. In fact, a recent survey found that almost one-fourth of all older adults now live more than 100 miles away from their children!

No matter which generation moves away, or for whatever reasons, living far apart changes how we see each other, as well as how often we see each other. It means that we cannot visit easily but, instead, our visits must be carefully planned. We must arrange time and rearrange schedules, pack belongings, prepare house space or housecare, pay expensive travel fares, or drive long distances. A visit becomes a big trip, as well as a "big deal."

Presumably, distance from our children does offer some advantages since, by not seeing each other so often, we do not wear out our welcome. In fact, we may be treated like special guests in each other's homes, and enjoy the pleasures of special favors and fuss. We may also be relieved of certain expectations that can come with proximity, such as for childcare or household assistance. The distance that separates us may also protect us from certain aspects of our children's lives that we do not need to know about, or do not want to know about. Yet, despite gaining something, our distance from our children feels more like we are missing, or losing, something.

When we do visit, we always seem to have either too much time, or not enough time, together. Because of the effort and cost involved, we tend to make extended visits, instead of brief, frequent ones. During these times, we are together constantly, intensely, and probably, excessively. Even with loved ones — perhaps especially with loved ones — so much togetherness can be a strain.

On the other hand, there never seems to be enough time to make up for all the long intervals apart. Since we are not part of our children's ongoing lives, with each visit, we feel we must repeatedly try to "catch up." Since we see each other only seldom, we want a great deal from each visit — probably too much; and after each one, we experience mixed reactions. We feel disappointment that the visit was not all we wanted it to be; we feel loss because we know it will be a long time until our next chance to be together; we feel resignation because we know the next visit will probably be just as imperfect as this one; and also, if we are willing to admit it, we feel some relief when the time comes to return to our own homes and lifestyles.

When I stay in my children's homes, I feel guilty about disrupting their households, changing their schedules, or causing them additional expense or labor. I have found that it is not easy to be a guest in my children's homes, no matter how hard they try to make me welcome, or how hard I try to accommodate their needs. It is not easy to live in their world, with its different experiences and expectations, especially if you see things you do not agree with, or approve of, or that worry you. It is also not easy, as parents and older persons, to accept someone else being in charge, even if it is our own children, and even if it is only temporary.

When my children come to my home for a visit, I immediately and totally rearrange my life. I shop and prepare and cook, to provide too much of everybody's favorite food; I spend too much money on special gifts and too much time on special treats. I try so hard and work so hard that I am usually overexcited when they come, and completely exhausted when they leave. It is not only the physical effort, but the emotional effort of exercising special vigilance to make certain that nothing spoils the visit. It is difficult to relax while I am trying to be so careful to avoid any reminders of old disputes, or any reasons for new ones. I cannot help but wonder whether, deep down, we all feel like strangers in each other's homes now, even though we are parents and children to each other.

Geography makes me a "long-distance grandparent," as well as a "long-distance parent." A recent survey of modern families found that the majority of children and their grandparents, approximately 80%, have only intermittent contact with each other due to geographic separation. So I am not alone in being so far from my grandchildren. But that is scant comfort.

I see my grandchildren a few times a year, when they come to visit, or when I visit them; but I cannot be a regular and natural part of their lives. I cannot be there to watch them in the process of their growing up, or be available to help them, as well as watch them. I also cannot be there to help my adult children with that process. I know this is a loss for me, but I also think it is a loss for them.

I envy the other set of grandparents, who live nearby, and who can be readily available and readily involved. I worry that my grandchildren will know these other grandparents better than they know me, and that they will like them more, merely because they see them more. I worry, when they are little, that they will not remember me from one visit to the next; and when they grow up, that they will not know me as a real person, but only as a part-time, long-distance relative.

The reality is that my adult children cannot change their lives to live nearer to me; they have established their own independent homes, families, and activities where they are. Neither can I change my life to move nearer to them, because I also have my own independent life where I am. I wish there could be some kind of optimal middle distance that would allow all of us to maintain easy intimacy without losing independence. I wish that my children and grandchildren were close enough for us to see each other frequently and easily, but distant enough to protect our privacy. I know that I want it both ways; I want each of us to have our own lives, but also to be part of each other's lives. I know that it is hard to measure what is near enough and what is far enough in our relations with our grown children, and it is not just geography that I mean.

Our Children's Second Time Around

Once upon a time, when our grown children married, we assumed that the marriage would last and that they would live, if not happily- ever-after, at least together-ever-after. That was how our own marriages turned out ——permanent, but not necessarily perfect — and that was what our generation was raised to expect.

So when our children marry, it is generally a joyous occasion. We feel a sense of fulfillment of our parental responsibility, and also a sense of relief. Our expectation of a happy ending is not just for our own gratification, but we genuinely believe that what worked for us will also do well for our children. We have hopes and visions of their happy lives, of watching our children and their spouses mature and succeed, and of enjoying our grandchildren. It is like a reward we expect, and feel we deserve, as parents; or like reliving, through our children, our memory, or our fantasy, of what it was like for our parents and ourselves a generation earlier.

In our generation, remarriage occurred only in later years, after the death of a partner ended the marriage. But now, in our children's generation, remarriage occurs in younger years, after divorce has ended the marriage. According to current statistics, almost one out of every three young marriages ends in divorce, and about half of these divorced persons eventually remarry. Given these numbers, it is not surprising that our children's marriages may be part of the statistics; and the chances are more than likely that their first time will not be their last.

If our children's marriages *do* come to an end, we don't know whether to applaud or to criticize, whether to feel relieved that they are free, or concerned that they are not sufficiently committed. When they are divorced, we worry about their being alone and not being able to find someone else to share their lives; but when they remarry, we worry about that, too!

We feel torn because we want them to remarry, but we also want reassurance that this time will be all right; and we know that such reassurance is not possible. We worry that our children, who have already suffered pain and disappointment, may be running new risks in a new marriage. We worry that they may be too cautious, or not cautious enough, in selecting a second mate, that they may seek impossible virtues, or settle for less than they deserve. We worry that the new partner, who is also most likely divorced, may be bringing his or her own past mistakes to the present marriage. We worry because we don't know whether the lessons of the first marriage will strain or strengthen the second one.

The first time our adult children married, we never even thought about the possibility of divorce; the second time, it is the main thing we think about. Our pleasure in the new marriage is shadowed by the memory of the failed marriage.

Our relationships with our in-law children are also different the second time around. It is usually harder to invest in the new relationship, so we tend to be more guarded and cautious, less completely comfortable or trusting. Of course, the converse is equally true of our new in-law children's feelings toward us. Although we all try to be willing, we are still wary of each other.

Having lost our sense of security, we become oversensitive to any signs of strain in the new marriage. Marital disputes or differences that we once might have dismissed, or assumed would be satisfactorily resolved, now cause us worry. Sometimes, we try to protect ourselves by not looking too closely or understanding too much, by accepting appearances and reassurances. But this is merely a form of denial, since what we don't know can't hurt us, and we can even pretend it does not exist. Each anniversary becomes a celebration, and also a relief. When well-meaning friends ask us about the new marriage, we often temper our responses with conditional phrases, like "as far as I know" or "from what I can tell so far."

In all the current concerns about divorce and remarriage, there has been little or no attention paid to effects on the *parents* of the couple. Obviously, we do not suffer the same direct pain, disappointment, and disruption that our divorcing children do. But, as their parents, we also suffer. We feel the loss of our power to help them, to solve their problems, or to protect them from pain. We also feel the loss of our hopes for their happiness, and the feeling of security about their lives.

Most of us have learned, by this age and stage of life, that not all dreams come true. Yet, while we do not mind so much having to accept losses or limits in our *own* lives, it is not what we wanted for our children's lives.

I suppose that we have to believe that our adult children have learned enough from the first experience to become wiser and stronger in the second one. Perhaps the ultimate consolation must be that these are *their* marriages and *their* lives; that it is *their* hopes and needs that must be met, not *ours*. Yet, despite such acceptance, we cannot help envying those of our peers who enjoy their children's stable marriages and secure lives.

When the Bough Breaks, the Cradle will Fall

Whether we want it to be or not, parenthood is a lifetime, not a part-time, position, without any time off for good behavior (either our own or our children's!). No matter how old our children are (they may have children, or even grandchildren, of their own), we never stop feeling parental. By that, I mean not just caring about them, but continuing to feel responsible for what happens to them, especially for their unhappiness.

Somehow, we seem to feel that, as their parents, we should have been able to protect them from all pain, and to save them from all suffering. We blame ourselves for their sorrows — if we had only been wise enough or willing enough; or if we had only made the "right" decisions (whatever those may have been); or if we had loved them differently (perhaps more, or perhaps, even less); if we had just been better parents, then our children would have had better lives. Judith Viorst calls this the parental myth of "the guardian angel and shield of invulnerability."

We are certain that we would not do anything knowingly to harm our children, but what about unknowingly? We fear that our flaws as people, not just as parents, could have, in spite of our best intentions, contributed to our children's pains and problems. Therefore, no matter what damage afflicts them, we find ourselves somehow at fault.

All of us, as parents, held certain hopes for our children's lives when they grew up. We hoped they would have lasting and loving marriages, family and financial security, good health, and good humor. Not surprisingly, our dreams for them were patterned after the positives in our own lives. Problems were expected to occur, but were also expected to be manageable.

Instead, so much seems to go wrong that could not even be imagined, let alone managed. Their marriages flounder or fail; their jobs dead-end or break down; they suffer stress symptoms and instability; their growing children — our grandchildren — cause them grief when all of their efforts seem to be frustrated, or fall apart. Even if what they suffer are disappointments, and not really disasters, even if they are trials, not tragedies, this is not what we worked for, or wanted for them. Even though these may be due to circumstances beyond our control, we still feel responsible; even though these may be due to our children's own choices, we still feel remiss.

Yet, don't we remember, when we look back at our own younger selves, how we were directed by our own desires and decisions; how we needed to pursue our own paths, not our parents' paths? The truth of the matter is that our children are finished with being parented by us long before we feel finished with being parents to them. In other words, we cannot claim either full credit or full blame.

Nevertheless, when these dreams for our children don't materialize, we suffer on two levels: we share the losses in our children's lives; and we also feel the loss of our own power as their parents. At least, when our children were small, we knew we could help them. When they were hurt, we comforted them; when they were in danger, we protected them; when they were sick, we healed them. We could put Bandaids on their scrapes, mend their dolls and bicycles, and chase away the goblins in the dark. But now, as adults, they suffer heartaches, not stomach aches; and it is their lives, not their toys, that are broken.

What can we do as parents now, when there is so little we can do, in parenting roles, for them anymore? The things that we can still do for them don't seem to matter anymore, and the things that do matter, we can do little about. We can no longer give them instructions, direct their decisions, control their environments, or smooth their paths. All we can give is advice (and even that sparingly and selectively) and the certainty of our support. Yet we still feel the desire to parent our grown children, even though we know we have lost the capacity and opportunity to do so. It is hard to accept the fact that we can no longer "save" our children, and that we can now only do more hoping than helping.

We also must let go of our own hope of reward for being "perfect" parents. It is not that we were motivated by repayment but, deep down, I think we did expect some kind of reward or, at least, some kind of satisfaction for a job well-done. I think we expected from our grown children the same kind of gratification we remember giving (or thought we gave) to our parents when we grew up. Especially now, in our senior years, we looked forward to enjoying peace and pleasure with our adult children. We did not expect, at this late stage, that the cares might continue, or that the worries might worsen.

Our sense of ourselves as parents seems to depend so much upon how our children's lives turn out; their successes validate us, and their sorrows accuse us. So any reward we receive is mixed with regret whenever our children's lives turn out to be more difficult, or different, than we had hoped for them.

If, indeed, we let go of our hopes for our children's lives, of our sense of power as their parents, and of our own dreams of reward, then what remains? Mainly, we are left with the hope that life will be kind to our children and that, when it is not, they may have learned enough from our years of love and our years of lessons to cope. I suppose that is why being parents of adult children feels both so powerful and so powerless!

Our Children's Sibling Relationships

Lately, I have begun to think a lot about my children's sibling relationship with each other. In the past, I thought about them more in terms of being my daughters than of being each other's siblings; and I simply assumed that the relationship between them would continue, virtually unchanged, indefinitely. But I have now become increasingly aware that, at some time, I will no longer be available to continue as the family "kin-keeper." When that time comes, my children will have to move on from being daughters to their parents, to becoming sisters to each other.

You would think that you could just assume there would be a close relationship between your children; that, because they were born of the same parents and raised in the same home, they would grow up like each other, and liking each other. But we have all seen too many examples of adult sibling estrangement to take this for granted. It is ironic to note that when we want to describe a particularly close relationship with someone, we say, "We're just like sisters," or, "He's just like a brother to me." But when we say this, we are comparing with a wished-for, not a real, relationship, because the truth is that siblings are not necessarily always, or automatically, close to each other.

With my own two daughters, I didn't worry so much about their being estranged because they disliked each other, but because they were so different from each other. The fact that they were so different was often a puzzlement to me. After all, I wanted the same things for both of them, and I thought I did the same things for both of them. So how did they become so different, when I tried so hard to make things the same? It is interesting that, although they were not alike and never looked alike, there was always something about them — their style, their speech, their mannerisms — that immediately marked them as sisters, even to strangers seeing them for the first time.

When they grew up, they moved down different paths in their interests, in their careers, and in the men they married; in the ways they raised their children, and even in the ways they behaved with us. It wasn't that they quarreled with each other; they simply had little to do with each other. It wasn't that they disapproved of one another's lifestyles, but that they couldn't relate to them. It wasn't that they wouldn't, or didn't, help each other in times of need, but they didn't really understand each other's needs.

As their mother, I was certainly aware that their relationship was not always the close one I wanted it to be, or fantasized it could be. So I tried to bring them closer together by patching up differences that I could not dismiss, or by soothing and smoothing ruffled feelings. When I saw misunderstandings or disagreements, I tried to mediate by explaining and defending one to the other.

I know that adult sibling relationships are often shaped by memories of childhood experiences, real or imagined; and those experiences, in turn, are often shaped by real or imagined comparisons of treatment by their parents. That means that, as their parent, I carry responsibility not just for the relationship between each of my daughters and me, but also for the relationship between each of them.

As far as I can remember, I tried to treat my two daughters as equally as I could; and if perhaps my treatment of them was not always equal, at least I was as fair as I could be to each of them. Though each one was my child, each one was a different child, so that meant I had to be a different mother to each of them; but being different was never intended as being less, for either of them.

Now that I feel more urgency about my children's relationship with each other, I also feel less power to do anything about it. At this stage, I can no longer intervene or interpret for them; I cannot defend or define them anymore. I cannot correct any mistakes I made in raising them, or retroactively, apply wisdom now that I did not have then. I always wanted my children to be close to each other, as well as to me; but I realize now that such a connection must be their own doing, not assigned to, or designed for them, by others — not even by me!

And now I have begun to notice that, without any hint or help from me, my daughters' relationship with one another is beginning to change. It's as if they have stepped across a new threshold which has moved them into new sibling roles. They have begun to do more things together and consult together more; they have more telephone conversations and share more ideas; they visit with each other, even when there are no special family occasions or crises requiring their attention.

Perhaps raising children of their own has brought them closer together. Now they share a common role as parents themselves that is distinct from their past roles, and past differences, as daughters. Perhaps as they watch their own children grow up, they can better understand how it is that their parents have loved them equally well, though necessarily differently.

Or perhaps, even though they do not say so to us, they share common concerns about my husband and me as they watch us begin to age. Perhaps they have begun to see us as not only shaped by strength, but shadowed by mortality. Perhaps they feel a need to draw closer to each other as they realize that, in time, we will no longer be with them, and they will be the ones who are left with each other. Finally, it seems that, despite whatever has been so different between them, they have now become aware of what is so similar.

Our Children, the "Baby Boomers"

We hear a great deal lately about the "Baby Boomer" generation, especially now that the first of them are beginning to turn 50. This is the group that was born right after the end of World War II, and which comprises the most numerous generation in our history.

As these Baby Boomers begin to move toward middle age, the various experts — sociologists, economists, and demographers — have begun to worry about their impact on our society. They worry about the financial drain on the Social Security and Medicare programs; about the strain on limited available resources; about the effect on the work force and the economy. There are even some who darkly predict competition between this newly aging middle generation and the already aging older generation, because too many Baby Boomers will be needing the resources which too many older people are using up by living too long!

But for us, these 40-ish-turning-50 Baby Boomers are not some threatening or competing army of strangers; they are our children! They are the children we had when the men came home safely from World War II, and we could then begin to plan families and futures for ourselves. They are the children we tended through their infancy, suffered with through their adolescence, and tried to guide into their young adulthood.

That is why it seems so strange to hear them described and categorized now as the new middle-aged generation. How can they be middle-aged when they are still our children? How can we think of them as no longer being the "young generation" when they seem so much younger than we remember being at their age? When we were entering middle age, we were more settled and certain, perhaps even more stodgy. We were someone's parent, not someone's child. Has it taken our children longer to grow up than it did for our generation, so that their middle age is really different than ours was? Or is it just the way we still see them?

But, at the same time as they seem too young to be middle-aged, we are shocked to see the signs of their actually becoming middle-aged. We notice their first visible gray hairs; we hear our daughters talk about pre-menopausal symptoms and our sons talk about prostate problems; we listen to them discuss their pre-retirement planning. One writer described the Baby Boomers as " ... now entering a different land ... that of their fathers and mothers." It is hard to believe that, in only a few more years, they will be eligible for some of the same senior discounts and organizations that we are! And if they are beginning to age, what does that mean about us, and what we are beginning to become?

Not only do we remember so recently raising these Baby Boomer children, but we also remember so recently remonstrating with them. For these were children who brought us great difficulties, as well as great delights. We differed with them over their taste in music, and clothes, and hairstyles; we argued with them about drugs and pre-marital sex; we quarreled over values and Vietnam. We wondered whether they would ever really grow up, and now we wonder at how grown up they have become.

We see our once-rebellious children leading adult lives that begin to resemble our own in many ways. Now they are the ones who worry about the cost of college tuition, about job security and family support, about savings and investments — all the sober responsibilities about which we used to be the ones to worry. Now they argue and remonstrate more with their own children, while they argue less with us, their parents.

Not only that, but they are even beginning to worry about some of the same things we do now — health, crime, and safety — not about changing the world, but about coping with it. It used to be that when they thought about such things, it was out of concern for us. Now, it is concern for themselves, as well.

In a recent issue of a national magazine, a major article featured interviews with a number of Baby Boomers. Virtually all of them expressed feelings of appreciation for their parents — about their sense of responsibility and realism, about their hard work and commitment — sometimes, much to their surprise, and despite any continuing differences they may still have with them. But it works both ways, because we too are learning from our children — from their independence, their individualism, and their idealism — also much to our surprise, and despite any differences we still feel with them.

Perhaps it was not really possible for us to appreciate each other before this. Perhaps we can understand each other better now, because they have more experiences, and we have fewer expectations. Perhaps it is because neither of us has the need to prove anything to each other anymore, so we can settle for accepting each other.

I think that both of our generations share mixed emotions about what is happening to us, with varying degrees of relief, grief, and disbelief. We feel relief because the stresses between us are being reduced, and the resemblances between us are being recognized. We feel some grief, because we know it is the end of their youth, and the beginning of our real aging. And we feel a certain disbelief at how all this has happened, before we expected it, and before we were prepared for it.

Even though the Baby Boomers are no longer the youthful generation, they are the most youthful middle-aged generation in our society's history; but we are the most youthful older generation, as well. This means that we may have many years remaining to live with each other. I think that during these years, when we are past the stage when we need to parent our children, and not yet at the stage where they think they need to parent us, we can become closer to one another than we have ever been, and perhaps, can ever be again.

So, I believe that the so-called experts, with their dramatic descriptions and dire predictions, are mistaken. The Baby Boomers are not our rivals or our replacements; they are not our enemies or our opposites; they are not some unknown or disruptive strangers. They are simply our children, now grown up, and finally, grown closer to us.

Our "Real" Children and
Our In-Law Children

When our children grow up and marry, they bring a new person and a new relationship into our lives, as well as into theirs. Although by this time, if those marriages have lasted, we have known our sons- or daughters-in-law for 15, 20 or more years, that continued relationship is still complicated. Even the very term "in-law" is confusing. (How can a child be a son or daughter by law, rather than by birth?) How can a parent be a mother or father to a strange, grown-up "child"?

As parents, we wanted our children to marry, and we probably had our own visions of the kinds of mates we wanted them to marry. However, we probably also felt that, whoever our children married — no matter how well they met our expectations — they could not possibly be all that we wanted for our children, or all that our children deserved. We also wanted our children to choose spouses who would fit in well with our family, without in any way changing the family. But we also knew and feared that the entry of these in-law children would be both a gain and a loss for us. Deep down, we knew that the people our children married —even though we wanted them to do so — would take them away from us.

The reality of our relationship with in-law children is usually different from either our fears or our fantasies, and it is also different later in the relationship than it was at the beginning. At the start, we are not even certain about what to *call* each other, let alone how to *behave* with each other. Should they call us "Mother" and "Dad," as if we are equivalents of their real parents? Or should they call us by our first names, as if we are simply older friends? Should we refer to them as our "children," our "children-in-law," or by no special label at all?

We are wary of each other; cordial, but careful about what we say or do in each other's presence. As parents, we make special efforts to remember their birthdays and accommodate their particular tastes and interests. But this takes reminders, since they are not the same, natural responses we have with our own children. As the younger generation, they are invariably polite and correct with us, but we still sense some of their distance and discomfort. We tend to interact with our in-law children more through our own children than directly with one another, so we do not get to know one another too easily, or too well. Of course, much of this depends on our relationship with our own children; in some instances, an in-law can become the "good" child who makes up for problems with our "real" child, or can become the "bad" one, who is blamed for those problems.

Over the years, mother-in-law jokes or cartoons notwithstanding, and no matter how we may feel about each other, we usually come to terms with one another. First of all, given the current divorce rates, we are grateful that our children's marriages have been lasting and loving. Secondly, we are grateful for the grandchildren they have given us. And finally, although this is difficult for us to acknowledge, we need to safeguard our relationship with our children and their spouses because we know, as we grow older, we may need them someday.

Consequently, we see ourselves as accepting of our in-law children, which may mean overlooking things we don't like and managing without things we would like, but do not have, with them. What we probably don't realize is that our sons- or daughters-in-law most likely see themselves in the same light in their relationship with us. Probably, we are both right; probably, we both make allowances for each other because we each realize that we have more to gain than to lose by getting along. Also, by this time, our in-law children have most likely learned that our interest in their lives is not necessarily an intrusion, but simply our only way of trying to remain parents.

Time changes all of us over the years. We are now no longer the intimidating, powerful elders we were when our in-law children first met us; and they are no longer the young, diffident newcomers who were so eager to be accepted. It's hard to believe, but they are now almost the same age, and almost at the same stage of life, as we were when we first met each other.

Our in-law children can never be the same to us as our real children —no matter how frequently and genuinely we proclaim our affection — just as they can never feel toward us the same connection they do with their own parents. When we exult that a daughter- or son-in-law is "just like" our own daughter or son, we are actually acknowledging that "just like" may be the most, or best, we can expect; but it is still not the same. Although it is a different and more limited relationship, in some respects, it may actually be an easier one. It is less emotionally invested and, therefore, has less power to bring us the kind of joy, or cause us the same kind of pain, that our own children can. Our relationship with them is important in our lives because it is important in our children's lives; and though we may not love them as much, we may sometimes enjoy them even more.

Part V

Feeling Our Age

Aches and Pains and Odds and Ends

There is a common notion that age and illness go together; sort of like love and marriage, "you can't have one without the other!" Being old is equated with being sick, and the common image of the aging person is one who is white-haired, weak, and worn out.

But the statistics, as well as observable evidence, contradict this notion. Older people today not only live longer, but they also remain stronger longer. It is estimated that approximately 80% of seniors are in moderate to good health, and only about 5% — usually the very elderly — are the frail figures of the stereotype. In fact, today's healthy 70-year-old is biologically equivalent to yesterday's typical 60-year-old. We all age differently, and there is no particular fixed year, or sign, which defines becoming old. In fact, there is a popular story about the late Justice Oliver Wendell Holmes who, in his 90s, admired an attractive young woman he saw and sighed, "Oh, to be 70 again!"

There is also a theory that women and men react to their biological aging differently, women being more adaptable to, and accepting of, bodily changes (supposedly, because of their built-in sensitivity to the life cycle). I think that the truth of the matter is much simpler than that. I think that we women seem to be less worried about our own health because, at this stage of life, we're too busy worrying about our husbands' health! (Perhaps it is an unconscious fear of widowhood.) At any rate, in tending to *their* symptoms, we tend to deflect our *own*.

As we face our aging process, the reality seems somewhere in between the statistics and the stereotypes. We know that, regardless of how well we feel or how well we take care of ourselves, we're certainly not going to become younger; if we're lucky, we'll just get older. Since we each age at our own rate and in our own way, the meaning of health is relative, depending on with whom, or with what stage of life, we are comparing ourselves. Not only does how we feel change as we grow older, but so does how we feel about how we feel. It gets to the point that being well basically means being well enough!

I think the first thing we notice is that our bodies can no longer do things they used to be able to do, or do them in the same way, or do them as effortlessly, as they once did. In the past, we probably weren't so conscious of our bodies because all of their parts functioned properly and automatically. Health has often been referred to as "the quiet asset." Now, because of interruptions or malfunctions of varying kinds and degrees, we become suddenly aware of our different body parts, which are no longer as reliable as they used to be. The changes are not necessarily drastic; in fact, they may be only small, but nagging, discomforts. But despite remedies and reassurances, we know that all these aches and pains and odds and ends that are happening to us are the beginning of a process that is irreversible.

We now need bifocals for seeing and aids for hearing. We use muscle relaxants, heating pads, and ointments for bursitis, arthritis, and neuritis. We take blood thinners and blood supplements. We use Maalox for indigestion, fiber for constipation, diuretics for edema, nitroglycerin for angina, insulin for diabetes, and Valium for tension. There is now too much room in the house, and not enough room in the medicine cabinet! We wear support hose for aching legs, orthopedic shoes for aching feet, and special corsets for aching backs. We're on special low-fat, low-sugar, low-salt, and low-cholesterol diets, which means that we must eat food that's good for us, rather than food that tastes good. We're now supposed to *watch* what we eat, not *enjoy* it! We're tired more during the day, and sleep less during the night. This stage of life has been described as being a time when your back goes out more than you do!

Another change I've noticed is that our reflexes are slower, and our reserves are reduced. It takes longer to recover from illness, infection, or fatigue. It's true that there are new medical/surgical procedures which attempt to replace what is being lost, like hip replacements, knee replacements, implants, and transplants. But this knowledge is small comfort when it seems that every body part we have either hurts, or simply doesn't work!

This means we have to make changes in our lifestyles, paying attention to diet, exercise, and rest — not for pleasure, but for preservation. We know that there have been exceptions like George Burns, who contradicted all the conventions and managed to stay fit and funny until he died at the age of 100! However, we enjoyed him mostly because we knew he was the exception, not the example. For most of us, it seems we practice not so much health care, but care of ill health!

There's a common expression that tells us we're only as old as we feel; but the truth is that we're really as old as we are. We need to understand, not to deny, our body changes. We know that we can only defer, not defeat, the aging process; so what we want is the longest possible period during which we can enjoy reasonable health and regular activity. In other words, we want "slow-motion" aging.

When I think about my own aging, I find myself alternating between reasonableness and rage. On the one hand, I am angry about my own sense of helplessness, and the feelings of unfairness that it engenders. After all, I'm certainly not ready to be written off yet! Now that I'm finally beginning to get it all together, I should not be falling apart! On the other hand, I recognize that there is a cost for living longer, and my body and I are willing to pay that price. After all, as French actor Maurice Chevalier once said, "Look at the alternative!"

A Prescription for Doctors

It seems to me that, when we get older and need to spend more time with doctors, the less time they seem to have to spend with us. When we were younger, we visited doctors only periodically for preventive check-ups, or occasionally for particular illnesses or injuries that were cured within a specific time, or with a specific treatment.

Now that we are older, we see doctors frequently and regularly for a continuum of complaints that seem to have no crisis points, nor any cure. We now need more doctors more often for more problems, even though we know there is less they can do about them. In fact, we see a great many different doctors — internists, cardiologists, neurologists, orthopedists, rheumatologists, and ophthalmologists — and take a great many different kinds of medications for all the different parts that hurt or don't work. Someone ruefully commented that his little black book of addresses now contains mainly names that end in "MD" Indeed, visiting doctors and filling prescriptions have become regular rituals in our retirement lives.

Adding to our discomfort, we know that we are not considered particularly successful, or even desirable, patients by our doctors. We do not provide them with opportunities for interesting illnesses or remarkable recoveries. We know that we are repetitious, and yet, may not remember the questions we are supposed to ask, or the answers we are supposed to give. We are often fearful and, sometimes, even foolish — but it isn't easy to be ill, and it's even harder to be older and ill.

Perhaps it is no wonder that so few doctors seem to have real interest and enthusiasm for treating elderly patients, despite the demographics that show us to be the fastest-growing population group and the one most in need of medical care. Those doctors who *do* treat us tend to alternate between two contrasting approaches: they either placate us with prescriptions, or they assault us with technology. In the first case, our medicine cabinets become filled, but our physical needs may remain wanting. In the second case, the various scans, probes, procedures, tests, or tubes may indeed make us forget our original complaints, because the new discomforts they cause feel even worse!

Sometimes our doctors try to comfort us with comments such as, "What can you expect at your age?" Frankly, I find that retort insensitive, because it makes me feel as if my aging is some incurable disease from which I have no right or reason to expect any relief.

In addition to all these changes in us, the patients, there have also been changes in modern medical practice. Most important is specialization, which means that medical skill has expanded as more is known, but has also contracted as it is applied more specifically. Each of our different doctors now knows more about each particular part of us, but less about all of us. The trouble with the new medical technology is that it has become more sophisticated and complicated than we are; although knowledge has become advanced and specialized, our needs have remained general and basic. While each of our particular parts may have particular problems, all of our parts and problems are connected.

There is also the matter of doctor's time. I have noticed that doctors are so busy now that we seem to spend more time *waiting* for them than actually *being* with them. First, we wait in the waiting room with other nervous patients. Then, we are finally led into a small examining room where we wait some more, but now alone, in varying states of undress and anxiety. The doctor stops at each examining cubicle as he works his way along the corridor, not waiting for anyone or wasting any time. Then there is the final, fearful wait for the doctor's return with the results and recommendations. It is certainly an efficient and time-saving arrangement for the doctor but, as the patient, I only wish that as much attention were paid to taking care of my *emotional* assurance as is paid to my medical insurance!

Another thing I have noticed is that the older I get, the younger the doctors seem to be. Doctors used to be comforting father figures; now they are young enough to be my children's age, and sometimes even look like my grown grandchildren. It is difficult to bare myself, body and soul, to these young people with unlined faces and barely lived lives. It is difficult to confess my fears and my body failings to someone who reminds me of my own grandchild. It is difficult to really request help from members of a generation so young that I was only recently helping them to dress!

There is also the matter of names. My generation was taught to respect our elders by addressing them by proper formal names and titles. Nowadays, our doctors (who may be *half* our age) call us immediately and casually by our first names. Even the nurse or receptionist, who may never have seen us before, feels free to do so, without permission or penalty. It makes us feel as if, by becoming patients, we are no longer entitled to the customary courtesies accorded our age. How I am called may not seem important medically, but it feels important to me personally.

I wish I had the power of decision over whether, or how, or when to be ill; but since I do not, the best I can do is make clear my preferences and prohibitions. First of all, I would start by responding to any of my doctors who ask me what I can expect at my age. I expect what any patient at any age expects from the medical profession — the best available professional knowledge exercised on my behalf; a reasonable amount of attention, patience and concern; and the safest and soonest possible relief. I would also appreciate a little respect for my age, without feeling that I must make excuses for it. I do not expect more treatment because I am older, but I will certainly not accept less! I am old enough now not to expect miracles or magic, but not old enough to accept mistakes or mistreatment. In other words, though I may not necessarily expect a cure, I *do* expect care!

Sensing the Difference

Old age has been described as a time when print becomes smaller, lights become dimmer, food tastes blander, roses hardly smell at all, and people are always mumbling. It certainly seems so to us, as we find our senses changing with age, along with the rest of us.

Our senses have been called our "windows on the world," the means by which we receive stimuli and perceive what is happening around us. They not only bring the world to us, but also provide us with the pleasures of seeing, hearing, tasting, smelling, and touching. But what happens when those "windows" begin to cloud up, or close down? What happens to the sensory pleasures we enjoy, and to the world we know? It is not that our senses change drastically or suddenly, except in cases of injury or accident, but they do change gradually, and inexorably, over time.

I know that sensory losses are expected at my age, and some of them are even correctable with special aids or devices. I am certainly grateful that there are such devices for me to use, but I regret having to use them. They make my impairments obvious and visible; they are mere substitutes for what I used to be able to do, and reminders that I can no longer do them. I also find myself more conscious of my senses, now that I can no longer take them for granted.

Vision is usually considered the first of our senses in importance, but it is also usually the first to start failing. Most older people wear some kind of glasses, at least some of the time, for some purpose; in fact, those who do not require glasses at all are the remarkable exception, not the rule.

I have been wearing glasses for so many years now that it is hard to remember myself without them. I started with glasses for close reading; then for writing, working and driving; and finally, just for seeing. I moved from plain reading glasses to bifocals; and may soon need trifocals. Glasses correct my vision, but they complicate my life, because I must now carry a virtual arsenal of equipment with me — one set of glasses for long-distance vision, another set for reading, or bifocals for either one; glasses for indoors, and prescription sunglasses for outdoors; along with different kinds of eye drops for dryness, soreness, or tiredness. I have always loved to read, but it is now becoming a slower and more tiring process — even without any of the serious vision problems that I know others have, like cataracts, glaucoma, or macular degeneration. I know that there are large print volumes available in the library for this when the time comes, but so far, denial (or vanity) have been delaying my use of those resources.

For some inexplicable reason, hearing impairment seems less acceptable than vision impairment, and wearing a hearing aid is more embarrassing and more avoided than wearing glasses. Somehow, a hearing aid, even more than glasses, seems to add to our years and subtract from our looks. Indeed, even those who are resigned to hearing aids have been known to frequently "forget" how to use them, or even to "forget" to wear them. Many of us simply deal with our hearing changes more through cover-ups than corrections.

For myself, I try to sit near the front at a movie or theater, and I increase the volume of the television set at home. I listen carefully when people are speaking, and I try to avoid conversations in large crowds, or large rooms, or places with background noises. Otherwise, I miss part of the sounds, and therefore, the meanings, of what is being said. The greatest frustration for me is when everyone else is laughing at something funny that I have not understood, because I did not hear it well, or hear it correctly. Finally, I have also learned to ask people to repeat what they have said, or to speak more slowly or loudly — without embarrassment or pathos — although that is not easy for me to do.

Another change I have noticed is in my sense of taste. I remember when I was younger, how much and how many different kinds of food I could eat and enjoy. But now I have many restrictions on my diet. I cannot have too much salt or too many spices; I cannot have fried or fancy foods; I cannot have sodium, sauces, or sweets; I cannot have too much caffeine, or too many calories; I cannot have large portions, or late portions. I am now on a diet that is low in cholesterol, low in preservatives, and low in taste. And all this is in addition to the natural weakening and wearing down of my taste buds as I grow older; so food just isn't as much fun anymore.

We tend to pay less attention to the gradual losses in our senses of smell and touch. Perhaps it doesn't seem like such deprivation not to have to suffer unpleasant odors, or not to have to feel scrapes or scratches as much as we once did. But on the other hand, it also means not being able to fully enjoy the special aromas of cooking, or the soft touch of silk on our fingertips. Even though these are perceived as lesser losses, like our other sensory changes, they are still losses.

Yet there are still many sensations left for me to enjoy. I can still read some interesting books, and listen to the rich sounds of music; I can still taste some of the food I like, smell some new spring roses, and feel the softness of a baby's cheek. Also, because my life is not so busy or crowded now, I actually have more time to "smell the roses." And conversely, because I am older and know I have less time, I take more pleasure in the senses I still have. So I suppose that, overall, it is a question of whether the glass is half-full or half-empty. Should I waste time missing and mourning what is gone, or should I try to enjoy what still remains?

Mind Over Matter

I listened to a lecture recently about mind and memory changes in our later years. The speaker, who was attempting to reassure those of us seniors in the audience, said that if we were still worrying about our mind or memory failing, we probably had nothing to worry about. I assume he meant that if we remember enough to know that we have forgotten something, our mind is still working relatively well; the real problem occurs only when we are not aware there is a problem.

He also cited statistics that only about 5% of older people actually suffer from some form of dementia, compared with the common forgetfulness that comes with aging; so the odds are quite favorable. The problem with these statistics is that dementia is such a progressive disease, how can I be sure that the memory losses I now experience are going to be the full extent of my memory loss, and not merely the beginning of the progression? The result of the speaker's intended reassurances was to leave me feeling strangely unreassured.

I expected that I would change and slow down when I got older, but I always thought of it mainly in physical or medical terms. Though I knew that my body would be bound to age, I did not realize that my mind would, too.

Lately, I have noticed that I don't think as fast as I used to, or as well as I used to. I have to pay exceptionally close attention to try to grasp new information, and I cannot always recall what I am told the first time. Sometimes, I cannot remember particular names or words, even though I know exactly who and what I mean. In those situations, I have found that it is virtually impossible to force recall, no matter how much I try. I just have to let it go, and some time later, somehow spontaneously, the memory returns — usually when it is no longer needed! I forget what stories I have already told others, so I tell them again — and again. I also forget what stories others have told *me*, and insist that I have never heard them before. When I am speaking to someone, any interruption or distraction can derail my train of thought, and I find it hard to return to where I started. I don't know which is worse, forgetting things, or worrying so much about forgetting things!

I know that some decline in reasoning and memory is normal and expected, especially after age 70, and does not necessarily mean major mental impairment; nor does it necessarily have to detract from one's life. In fact, in some ways, as we grow older, what we experience is change in, rather than loss of, mental capacity, meaning that we now actually think differently than we did when we were younger. We may have more difficulty with specific details, but we have greater ability with broader concepts. It is ironic that now that I can understand more, I can remember less!

I also know that there may be more to memory loss than mere memory, because we sometimes "forget" painful things in order to protect ourselves. On the other hand, we sometimes remember things, especially from the past, that surprise us and make us wonder how we can be so forgetful and so nostalgic at the same time!

Meanwhile, I have been learning different ways to cope, or even to cover up. I find that regular routines help me — doing things the same way, or at the same time, or in the same place — so that what may seem to be rigidity to others is really my own way of remembering. I have found different means of reminding myself, too — writing things down, rehearsing in advance, simplifying procedures, or staying with what is familiar. I have also learned not to compare myself to, or compete with, others, especially younger people, who can think more quickly and remember more easily than I now can. I have also found that a little laughter helps a lot, especially when my friends admit to similar memory lapses. We share self-deprecating jokes with each other as a way of comforting one another.

Yet I know that beneath the laughter, there is some sense of loss. I miss the way my mind used to work; I miss the way I used to be able to learn and retain new information accurately and effortlessly; I miss the way I used to be able to answer questions quickly and make decisions unhesitatingly. It is not so much the particular incidents of forgetfulness that are so troubling, because these are usually more embarrassing, or inconvenient, than incapacitating. What is upsetting is the way they make us *feel* about ourselves, and what we *fear* they foretell about our future.

I realize that aging is both psychological and biological, and I am learning to cope with a little of each kind of loss. Sometimes, I feel like the anonymous poet who wrote in a letter to a friend:

"Just a line to say I'm living and not among the dead,
Though I'm getting more forgetful and mixed up in the head.

I'm used to my arthritis and to dentures am resigned.
I even manage my bifocals — but I surely miss my mind!"

Weathering the "Big Chill"

Regarding the health problems accompanying aging, I have learned that it is easy to be brave when there is little danger, or to be calm when there is little cause for concern. Over the years, we become resigned to changes in our health. Our bodies hurt more, and don't work as well. But although we are able to accept inconveniences, we do not really expect incapacities. Serious illness is never easy; but serious illness in our older years, though more common-place, is even harder. It's been said that sickness is an aging process in itself; therefore, when we become ill as we become older, both our illness and our aging are increased.

The first reaction to a serious health threat is usually a sense of disbelief. I remember my own sense of unreality because my health had previously been so "normal" and unremarkable that I used to half-complain and half-boast, it was "boring." Therefore, this new frailty had to be some kind of mistake or misunderstanding, some kind of strange test, or trick.

Disbelief is followed by a feeling of powerlessness, because things are happening inside our bodies that we cannot see, understand, or control. We are not accustomed to having to worry about whether our bodies will work the way they are supposed to, because they always have. Now, these once-healthy selves have become more like threatening strangers.

The third and final reaction is fear, but there may be several stages of denial before that sets in. Since there are usually so many arrangements to take care of — X-rays, tests, probes, prescriptions, and preparations — that we can forget or deny problems by becoming immersed in specific details. The hardest times are the empty spaces in between doing something, or having something done; those times of waiting become times of worrying.

Or we may find ourselves alternating between wanting to be fully informed (in order to be fully prepared for whatever might happen) and conversely, wanting to be protected as much as possible from knowing about whatever might happen. Beneath the different kinds of denial we manage to devise for ourselves, we are really trying to find a frame of mind to help us handle what is happening. Eventually, we settle for a kind of numb reality, in which we realize how much all of our previously brave talk about mortality was only a theoretical exercise.

Even afterward, when our illness is seemingly all over, it really is not. No matter how well we recover, no matter how reassuring the results, no matter how healed the scars or stitches, we do not automatically return to the way we used to be. We may feel a sense of relief, or rescue, but rarely complete restoration. At our age, we know that we recover from sickness more slowly and less certainly, and we continue to experience varying degrees of residual discomforts, like aftershocks following an earthquake.

We also notice other changes, not just in how we feel, but in how we *feel* about how we feel. We become more aware of, and more worried about, our various body parts and functions. We become concerned if our appetite or energy slows down or speeds up; or if we sleep too much or too little, or too lightly or too heavily. We examine every lump or laceration; we think about every twinge of pain, no matter how brief or infrequent. It is more the recurrence of peril than of pain that we actually fear. Now we worry that behind every minor symptom, there may be a major problem; and even behind the apparent absence of symptoms, there may still be some unforeseen dangers.

We keep on monitoring ourselves, carefully watching for symptoms of dysfunction, although we really do not know *what* we are watching *for*. We worry that we are becoming hypochondriacs, or that others will think we are. So we sometimes pretend not to be concerned, although we always are, even if secretly and silently. We didn't used to be that way; it used to be our assumption that everything was fine, because it had always been so. But now, we have experienced a different truth about our health and about ourselves.

We know that we may be cured, but we are still not free of care. In my case, I was lucky. I remember being reassured that I would soon be "as good as new." But the truth is that, at our age, we cannot be "new" again; and though we may seem as "good" as before, we know that we are not the same as before. It is not just the formerly healthy body that is missed, but the feeling of invulnerability and confidence it provided. Now that we know how vulnerable our bodies can actually be, we can never again feel safe in the same way.

Supposedly, there are positive lessons to be learned from serious illness. We may be finally frightened into doing what we should have been doing all along — taking better care of ourselves. Or the experience may enable us to better understand the feelings of others who suffer. Or we may find ourselves more fully appreciating the time and energy we have remaining.

But deep down, we feel that we have lost more from our illness than we have learned. We feel this sense of loss even more when we are older, because we know that we have less time and less energy left for reconstitution. Even with "successful" outcomes, success is defined not by how much has been gained, but by how little has been lost. We know that no matter how seemingly cleared or cured of illness we may be, we do not come out of the experience the same as we entered it.

A Hospital is No Place Like Home

Being hospitalized occurs more frequently when we are older, but no matter how many times it happens, we never really become accustomed to it. Repeated admissions may make us more familiar with certain procedures, and even with some personnel, but this familiarity does not ease our anxieties. We know that not everyone comes out of the hospital cured, especially older, repeat patients. Some of us come out with serious or incurable conditions; and some do not come out at all. So we suffer the silent, usually unacknowledged, fear that on one of our now more frequent hospital stays, the outcome may not be a "happy ending."

One of the first things that happens to us when we enter the hospital is that we have to turn over total control of our lives, to total strangers. Furthermore, since hospital staff shifts daily, sometimes even several times each day, it means that those now in charge of our lives are constantly changing. We are controlled by an ever-changing hierarchy of aides who tend to our practical needs; nurses who provide medication and periodic instruction; and doctors who offer brief, official pronouncements. We are willing to submit ourselves to all of them because, being so vulnerable, we need to believe that they know what they are doing, and that what they are doing can help us.

We are handled in the hospital mostly by people who are virtually young enough to be our grandchildren, but who treat us as if *we* were the children, instead. We are stripped, bathed, fed, and led by them in acts intended to be caretaking, but that feel to us to be infantilizing. We revert, or are forced to revert, to a state of dependence that we outgrew long ago, along with our childhoods. Supposedly, dependence — the so-called "sick role" — offers the compensation of absence of responsibility. But for us, as older people, this is no comfort, because it points not just to our past state of childhood, but also to our possible future state of helpless old age.

In the hospital, we experience not only acute discomfort, but also acute indignity. We are required to request assistance or permission to perform simple personal or biological functions. We lose, or are compelled to let go of, our sense of modesty and personal privacy. Our body intakes and outputs are monitored and measured, and our most intimate parts and practices are subjects of study and discussion. As older persons, we became accustomed, in the past, to being given respect for our modesty; but now, as older patients, we find that our private parts have become public domain.

Another thing we notice is how we are kept carefully uninformed about our own conditions! Our vital signs are taken regularly and recorded on our charts, along with the results of our tests, our laboratory work, and our X-rays. All of the important facts and figures about us are entered onto the chart, which we never get to see! I realize that there is a rationale for this secrecy; the patient may not be able to decipher the medical vocabulary, or may misinterpret the information and become confused or concerned. In other words, the secrecy is supposedly for our own good! But the truth of the matter is that, as mature adults accustomed to coping with problems and controlling our lives, the less we know, the more we worry; and the more we feel we understand, the less we fear.

Our sense of ourselves changes when we are in the hospital, because the whole focus of our lives changes; in fact, it almost completely reverses. As older adults, we have valued our competence and our ability to cope; but now, the entire focus of attention is on what is *wrong* with us, not what is *right* with us; on what is failing, not what is functioning. In the hospital, it seems that the less we think or do for ourselves, the better we are regarded as patients. I realize that this happens to younger patients, as well; but for us, the reversal is even more unnatural and uncomfortable.

Our sense of time is also different in the hospital; time feels both filled-up and empty there, depending on the time of the day. In the daytime, we are kept busy with all kinds of housekeeping procedures. There is bathing time, linen-changing time, meal time, shift-changing time, and visiting time; along with the times set aside for our own personal medical procedures, like medication time, therapy time, and examination time. The hours of the day are filled with prescribed procedures; but the hours of the night are long and empty. As older people, we have grown very sensitive to the foreshortening and the use of the remaining time in our lives. But in the hospital, our time is not only used up, but is no longer ours; it is under the control, and at the convenience, of others. Yet ironically, because we are older, we are expected to be more patient, not less, and to care less about the loss of our time.

Finally, the time does come when, weak, but well enough, we are able to go home. We are grateful that the hospital was able to help us, but we are even more grateful that we are able to leave it. In the hospital, with all its resources for care and cure, we never feel as safe as we do in our own homes, on our own schedules, and under our own control. Obviously, we cannot prevent, or even predict, health problems; so there may be no way to avoid further hospitalization at some time again in our older years.

But having been hospitalized, we know that it is only when we are home again that we really feel well, and really like ourselves again.

Long-Term Care, Long-Term Fear

The family life cycle used to be more limited and predictable than it is now. Our parents took care of us until we were old enough to care for ourselves; then we married and took care of our own children; and by the time they grew up, our elderly parents had died. There was little need to worry about long-term care, because few lived long enough to need it.

Today, advances in medical technology have made it possible to maintain life much longer, even if at a lower level, making it more likely that some kind of care, like help with dressing, eating, walking, bathing, or taking medication, will eventually be needed. In fact, the longer we live, the more different, and difficult, kinds of care we will need. In addition, because we are all living longer, our expected caretaking relatives are themselves becoming senior citizens facing their own aging needs. These days, we have 70-year-old "children" responsible for their 90-year-old parents!

Caring for aged parents is not just a labor of love; it also hard labor, which may be beyond the capacity of individual family members to provide, no matter how much they would like to. As a result, the very elderly, at the very end of their lives, are more and more often being cared for in nursing homes. This may not be so much a reflection of family neglect as of family necessity.

Nursing homes relieve families of functional, but not emotional, responsibility. We still feel the pain of "losing" elderly parents while they are still living; and we still feel the guilt of being unable to take care of them the way they took care of us. We also feel the fear of foreseeing our own future in what is happening to them.

According to statistics, more than one million elderly people today require some kind of long-term care. By the year 2000, that number may be as many as six million! Although only 5% of the elderly require such care, usually, at the very end of their lives, that percentage is expected to increase to as much as 25%, because the over-80 population is the fastest-growing age group in our country today. As a result, we now worry about which of our loved ones, or even whether we ourselves, will be among those numbers. Indeed, long-term care has become the personal and financial shadow that looms over our last years.

Besides its emotional toll, long-term care carries a high financial cost. Over a period of a few years, sometimes only months, it can wipe out a lifetime of savings! Money that was so carefully and lovingly saved to provide gifts and memories for children and grandchildren must instead be handed over to professional strangers. After personal savings have been used up, the financial burden falls on family members, or on public assistance. In either case, it is not the way we wanted, either for our loved ones or for ourselves, to end our lives or to spend our savings.

Sometimes, I find myself feeling caught in a kind of time warp, because it was not long ago that *I* was the adult child worrying about my *parents* in nursing homes. Now, I wonder how soon I will be the elderly parent for my children to worry about. Both of my parents ended their long lives in nursing homes. For each of them, I was part of the decision-making, selection, placement, and monitoring process. Although I could not actually take care of them, I tried to be there for them. I visited them almost daily, to make certain they were well cared-for. I suppose I did this for myself, as well as for them, because it served as a substitute for my care.

My own two daughters live 3000 miles away, so I know that they will not be able to take care of me, no matter how much they care about me. Anyway, I don't want to place on either of them the burden of having to provide care for me, or the guilt of not being able to do so. I suppose I just want to know that, if they could, they would be there for me (as I was for them when they were young, and as I was for my parents when they were old!).

It is ironic that, as we grow *so* much older these days with the aid of the marvels of medicine, nutrition, and technology, we find ourselves worrying less about not living long enough, and more about living too long! None of us wants to see our loved ones end up in nursing homes; nor do we wish for ourselves to end up there. My own fantasy of aging was some kind of endless, indeterminate middle age, in which I grew somewhat more white-haired, and somewhat slower and frailer, but basically, stayed almost the same. In that fantasy, life just ended on that kind of note, without any infirmity or indignity in between.

Unfortunately, we do not have such control over our life processes, and we have little choice about when we grow old, or how we grow old, or about how long we will be old. Statistics assure us that we are remaining younger longer and becoming older and sicker later; so we can expect to continue reasonably well until we are in our 80s. But since medical technology enables us to survive beyond that, it means that we must get older, even if it is not until later. A popular current expression, "successful aging," is often used to describe our goal for our later years; but success cannot mean total escape. It can only mean final years in which our aging changes are as lenient, as late, and as livable as possible.

Part VI

Endings and Losses

Caring and Caretaking

When we married, we expected to care about each other; we did not expect that a time might come when we would have to take care of each other. We knew we would grow older together, if we lived long enough and our marriage lasted long enough, and that we would probably experience some health problems, or other aging symptoms. But deep down, I think we really expected that we would remain almost the same, and that our relationship would remain almost the same.

Statistics tell us that one out of every five people over the age of 75 will suffer from some kind of chronic, debilitating, or dementing illness, and that the percentage will go up as our age goes up! We're assured that this is not part of normal aging, so it may not happen to us at all. But it happens so often, as we grow older and live longer, that the randomness and uncertainty of these statistics are more frightening than reassuring. They also tell us that usually, though the reasons are not known, it is the husband who needs the care, and the wife who becomes the caretaker.

I have watched as this has happened to some of my friends, neighbors, and acquaintances. As I look at those women friends who are now caretakers, and at their husbands who were my husband's friends, I visualize us in their places, and I feel pity for them; I also feel relief, as well as fear, for us.

I've noticed that the caretaking seems to take place in three stages. The first is the easiest, despite the initial shock. At this beginning time, there is still considerable hope and energy. The patient still seems reasonably himself so much of the time that one can almost believe that the diagnosis is simply wrong. Maybe this will be the exceptional situation; maybe he'll get better — or, at least, not worse. There hasn't been much time or change yet, so things are manageable, although more difficult. If it's no worse than this, one can cope, and keep hoping that it will get no worse.

In some ways, the second stage is the hardest, because it doesn't seem to have any clear boundaries or guidelines. The patient is changing noticeably now, almost as if he's slipping away, even as one watches him. Sometimes, there are still tantalizing glimpses of the person he used to be, and one dares again to hope a little. Then unpredictably, without warning or transition, he fades back into the lost stranger. We begin to feel that we are on an emotional roller-coaster. Sometimes, he even disappears physically, as well as emotionally; perhaps he wanders off, or gets lost; and then there's the fear of what might happen to him, and perhaps even anger at him for causing that fear.

He needs more help and physical care now; he needs a more unrelenting kind of vigilance. Expectations of his own ability to care for himself have to be lowered. But even as more and more must be done for him, it is increasingly clear that no matter how much is done or tried, it will not really help in the long- run. Ironically, it's sometimes hard to remember that he's ill because, on the outside, he still looks so much the same. In fact, it may be the caretaker who is the one who is changing and aging visibly, not him!

He cannot be left alone anymore, so the caretaker can never be alone, yet feels alone most of the time. Strangely enough, there are also rare moments that are very intimate and precious, of feeling very close and of almost enjoying the dependency (like a mother with a beloved child). But most of the time, there is rage against what is happening to him because, in a way, it is also happening to the one who cares for him; both are victims of the illness. And all the time, there is the wondering about how long the caretaking can go on.

The third stage comes finally, inevitably, and not easily. It comes when the caretaker can no longer handle the physical care, even with respite, or even with help. It comes when there is so little left of the person he used to be that it's as if he ceased to *exist*, but continues to live. Even then, the decision to "let go" is agonizing.

There is a final assault of questions and doubts. Is there such a thing as a right time for placement? Can one wait longer, or do more, or do something differently? What would he have done in this circumstance, had it been me who needed to be put into a home? What will our children, our family, or our friends think? Do they judge or sympathize, those others who have not themselves been the "hidden victims" of the illness? Do they understand the pain of letting go, even of someone who has, for all intents and purposes, already gone?

After the actual placement, there is a feeling of relief, but it is tinged with feelings of emptiness. In some ways, it is even more difficult than a death, because of the feeling of responsibility for making the choice. The caretaker is now neither wife nor widow. She feels envious of other couples who are healthy and together. She feels relief, mixed with guilt, for still being healthy herself. But sometimes, she also feels fear at recognizing similar symptoms in herself, such as memory lapses, or confusion, because it may be starting to happen to her, too.

Mostly, the caretaker visits as often as possible, although sometimes briefly, because there is no longer much to say. She monitors his care, making sure he's clean and shaved and fed (at least, that's something that she can still do for him). But sometimes, placement lasts so long that one caretaker I know called it an "endless funeral."

All this is what I've learned from watching what has happened to my friends. I realize how lucky we've been so far, my husband and I, as we and our marriage grow older. Does he ever wonder, as I do, whether it will happen to us? Does he ever wonder, as I do, whether it's worse to be the caretaker, or the one who must be cared for? Meanwhile, we feel grateful, but can never take for granted, that we will only need to care *about* each other, rather than have to take care *of* each other.

Losing One's Life Partner:
One is a Lonely Number

I am one of those fortunate older women who still enjoys a long-term marriage with a reasonably active, reasonably healthy mate. But I am repeatedly confronted with statistics which threaten otherwise. The experts warn that three-fourths of all wives will eventually be widows, outliving their husbands by seven to 10 years. Over half of the older women in America today are widows — approximately 10 million — with the number increasing each year.

I am further reminded because I live in a retirement community in which approximately 25% of the population consists of women alone. As I absorb the statistics, and as I observe the widowed women around me, I am made increasingly, and reluctantly, aware of the future that is possibly in store for me. I know that it has happened to all those other women and their husbands; but part of me cannot really believe it can happen to us — even though the alternative of my husband becoming a widower is equally unreal and unthinkable!

After many years of long-term marriage, we have finally learned to understand and accept each other. So it is ironic that now, when we need each other the most, we know that realistically, unless there is some major accident or natural disaster, one of us will die first, and the other will be left alone. I realize, with a certain shock, that I have never lived alone. Like most of the women in my generation, I went directly from the parental home to the marital home. I think about all the details of our life that my husband takes care of, and for which I need not be bothered or responsible. I think about all the things we share in the regular routine of our lives.

Who will do all the important practical things my husband always takes care of, if he's the one to die first? Who will take care of the house and car maintenance; the management of my finances, insurance, taxes, and investments? Who will protect me from danger — or, at least, make me feel protected?

Even though I have enjoyed intervals of quiet when my husband was away, what would it be like to have endless quiet? Will I go out by myself, travel by myself, dine out by myself? What will I do with all the lonely evenings? Who will compliment me on the way I look, or on the things I do? Who will there be to dress up for, or tell my jokes to, or argue with? Who will give me advice or feedback, even when I don't want it? Who will I share fun or fears with? Who will I laugh or cry with?

What will happen if I'm ill? Who will take me to the doctor, get the medication, keep me company, and provide consolation? Even if there is no illness, what will it be like to grow old alone, without companionship or comfort?

I know that my children care and would want to help, but how much should be expected of them? They both live far away and are busy with their own lives, and cannot be substitutes, or crutches, for my own emotional needs. The truth is that, despite whatever genuine loss they might feel, I would be the only one, not they, whose life would be totally altered.

Losing a long-term partner means losing the most important person in one's life. It means not only the loss of that particular, and precious, person in our lives, but also of a comfortable way of life. My friends who are widows tell me that they have had to move into a "different" social environment, meaning out of the couples' community, and into the world of women alone. It's not that our friends who are still in couples are uncaring, but it becomes awkward and complicated to deal with questions of seating and treating, dining and driving. The "left-over" woman is a social stress-inducer, not because of the threat of attracting other women's husbands (at our age, this notion is certainly more romance than reality), but because of the reminder of what has happened, and what may still happen.

In contrast, widowers are not viewed as social misfits; in fact, they are desirable social commodities. You don't read, or hear much concern, about a widower's need to adapt to this situation. I wonder if that's because there are so few of them, or because they don't have to learn to adapt, because they usually don't remain widowers for very long?

The older we get, the greater the likelihood of our being alone. As psychologist Eda Le Shan pointed out, the only companion we can be sure of for the rest of our lives is ourselves! Most of the time, I tend to alternate between fear and denial of this eventuality. I worry about my husband's health, but try to dismiss the thought of any real danger. I vacillate between opposites — learning to do independent things, as if in rehearsal or preparation for being alone; and not wanting know or do anything to prepare myself, as if to deny, or ward off, the threat of widowhood.

I ask myself questions I cannot answer: Should we spend more time together, or less? Should we try to adjust our relationship to be different with each other than we have been in the past, or should we just continue as we have for so many years till now, and for as many more years as we have together? I can only hope that if I am the one who is left, as statistics predict, I may have acquired sufficient strength and skill from our being together, to cope with our being apart.

Losing our Last Parent:
Becoming an Orphan

As long as we have one living parent, no matter how old, or in what condition, we are still someone's child. There is still someone to worry about us, to inquire about us, and to rejoice over us as only a parent can. There is still someone older than we are who is, in our memory or imagination, wiser and more powerful than we are. We can still retain the fantasy of them being parental and protective, although we know realistically that we no longer need either, and that they can no longer provide either. It may sound contradictory, but having even one parent alive makes us feel both more childlike (because we can still imagine turning to them to take care of us) and also, more adult (because we know that we can now take care of them).

In a sense, as long as our parents are still living, they are actually protecting us because it is they, not we, who are the older generation. It is they, not we, who are running out of time, and who are facing the fears of mortality. As long as we are still children to living parents, we are still the "younger," and therefore, the safer, generation. Worrying about their mortality spares us from worrying about our own.

When our first parent dies, the relationship with the surviving parent becomes more pressured, and also, more precious. First of all, there are the realities of the remaining parent's new needs. The surviving parent may now need assistance with certain tasks and responsibilities formerly shared with, or carried out by, the deceased spouse.

As their children, we feel the responsibility (and also, the genuine desire) to provide the help that is now missing. We cannot replace the lost parent, but we can replace some of the lost role responsibilities.

The remaining parent now receives a double share of our concern, attention, and affection. We spend more time with, and worry more about, this one. In fact, now that the relationship with the surviving parent is undivided and undistracted, we may actually get to know each other better, and feel closer to each other, than we ever did before. We feel more accepting of the surviving parent, even of those things that used to bother and burden us about them in the past. Perhaps the death of our first parent makes us aware of how vulnerable the other one is, and also of how limited our remaining time together may be.

It seems strange and surprising that the second death turns out to be even more painful than the first. It is true that there is less shock because we have already gone through one parental death; but, as columnist Ellen Goodman observed, "Death is always new," and even if expected, it is still never acceptable. The first time, there was still another parent left, so the feeling of still being a child to a parent — even though more symbolic than real — was also left. But after the second death, we are struck by the overwhelming totality and finality of our loss.

The longer our elderly parents live, the longer we are protected. I'm not certain whether that makes their eventual death easier or harder for us. During a period of three years, I watched each of my elderly parents die. I know that they both lived long and full lives, but is there really such a thing as a parent who is "old enough? " Is there really such a thing as being "ready" for your parents to die? When my second parent died a few years ago — he was already in his 90s, and I was already in my 60s — I remember that the first thought that struck me was that I was now an orphan!

Another surprising thing is that the better our relationship with our parents has been, the better able we are to cope with their loss. It's never easy, but it is easier, because at least we have the pleasure of positive memories to offset the pain. Even though we wish there could be more time together, we do not have the regret of wishing that it could have been different. On the other hand, if our relationship was conflicted, we know that now, those differences will never be remedied or resolved. Whatever was unfinished, unsaid, or undone is doomed to remain so. There will never be an opportunity to undo wrongs, to forgive and be forgiven, or to become the "beloved" child who finally enjoys parental pride and approval.

When our last parent dies, by virtue of that one single and irreversible event, we become both orphans and the older generation! Even our daily lives change in unexpected ways: The regular phone calls made to, or received by, our parents are no more; we no longer have to worry about seeing them, or seeing to their needs; we no longer have to live up to their expectations of us, or wonder whether they can meet our expectations of them. We now enjoy newly freed time for other things. But somehow, even with all this new "freedom," our lives seem as much emptier as fuller.

We realize that there are things now about our lives and our families that our parents will never get to share and enjoy. We have grandchildren whom our parents will never know, and who will never know them, except as we are able to reveal or remember of them. We even begin to worry that, over time, we will forget how they looked; and the fear of fading memory makes us feel as if we are losing them all over again.

When we bury our parents, we bury other things, as well. They were our strongest link to the past, to our own childhood, and to our childlike feelings; they were the last barrier between ourselves and our aging. We lose the final front line separating us from our own mortality when we lose our parents. An anonymous poem, entitled "The Orphan," describes it well:

"There's no umbrella now to separate you from eternity...
You're a survivor, with all the loneliness of survivorship ..."

Writer Chaim Potok stated that we cannot be truly adult in our own right until our parents are gone and we are released from their authority. Therefore, when our last parent dies, we become irrevocably grown-up. We realize that we have now taken our parents' place in the shifting of the generations. We also realize that we, too, are not forever; and that someday we, too, will leave our children, as our parents have left us. Then it will be our children who will feel adult and alone, as we do now.

When our Friends Start Failing

I remember that in our younger days, the state of our health was seldom a subject of conversation when we talked together with our friends. Our topics usually reflected the particular times of our lives: first, our marriages and young children; then, our business and professional accomplishments; later, our children's school and work triumphs; and finally, our beginning retirement plans, and the births of our grandchildren.

Our health was an invisible asset that was scarcely noticed because it worked so well, and therefore, was scarcely mentioned. Now, it is no longer a silent subject; in fact, it has become not only the core of our conversations, but also cause for concern. In the past, when we and our friends asked each other, "How are you?" we were referring to the latest news about happenings in our lives. Now, when we ask that question of each other, we all know that we are literally referring to the state of our health!

It used to be that it was the older generation, our parents and their peers, who worried about their health and their mortality. Indeed, that was one of the qualities that characterized the older generation and differentiated them from us. Serious illness held no great terror, or even meaning, for us at that time, because it did not seem to be part of our lives. I cannot remember when it was that we lost that feeling of invulnerability. We remember our parents in failing health, but when did we begin to take their place?

As I look around now, I see that it is the members of my own generation who supply the statistics on aging and illness. I see my friends suffering from, or worried about, a host of health problems, from hypertension to cataracts, coronaries, retinitis, rheumatoid arthritis, or slight or serious strokes. I see them undergoing bypass surgeries, mastectomies, and chemotherapy. "Remission" has become an achievement, and a successful health exam is a cause for celebration! I now find myself spending almost as much time visiting friends in hospitals, or making convalescent calls, as I do in social contacts; I now send out more "get well" cards than birthday or anniversary cards!

As I experience my friends failing, I also experience a mixture of feelings, for them and for myself, a combination of regret and relief, of frustration and fear. Even though they are the patients, I also feel part of their pain. I sorrow for what they are suffering, and I wish helplessly that it were not happening. But at the same time, I feel an involuntary, almost sneaking, sense of relief that it is happening to them and not to me. I may not want to feel relieved; I may even feel guilty about doing so; but I am not sorry that I have been spared.

I feel genuine concern for them and want to do something to assist and support them, but I am frustrated because there is so little immediate, or important, that I can do. At the same time, I also feel genuine concern and fear for myself. My friends and I are the same age and generation; we have always shared so much in common; will we have this fate in common, also? As a result, the dark clouds that now hang over them seem to shadow my life, as well.

I have also noticed that, as my friends' illnesses persist over time, I seem to go through a series of stages in my attitudes and actions toward them. At the beginning, I am almost over-involved, and overwhelm them with attempts to help, with frequent visits and telephone calls, offers of errands and favors, and suggestions and consolations. But after awhile, a gradual distancing and detachment begin to occur. The visits become less frequent, and the phone calls briefer; the urge to help and comfort diminish; even in our conversations, I want to talk about, and hear about, their sufferings and symptoms less. It is not that I become unkind or uncaring, but I become more uncomfortable with the continuing preoccupation with their problems. It is as if, at the same time, I feel conflicting urges to both be involved in their illness, and to escape from it!

It also seems to me that when our friends become ill, it is not only our conversations, but our relationships which change. Indeed, our friendships become another casualty of the illness. My friends who are patients are self-absorbed with their health problems, which are now the most important thing in *their* lives — but not in mine! Even with close friends, with whom I used to talk animatedly for hours, I now struggle to sustain conversations once the usual clichés of concern, even if genuine, have been expressed. Not only can we no longer do things together; now we even run out of things to say to each other. I miss what my friends used to be like, and I miss what our friendships used to be like.

In the past, when it was members of the generation older than ourselves — our relatives, teachers, and employers — who suffered serious illness, it was expected and accepted. Even when it was our own parents who failed, we sorrowed, but we still felt "safe." But now, it is our peers who are at risk, and therefore, by extension and association, ourselves.

Even though I may still feel in reasonable health and have not seen any major changes in myself, I see the changes in my friends. I know that whatever is happening to them may, and can soon, happen to me. I know that their problems may mirror and predict my own future. I can, and often do, try to pretend and protect myself by poking fun at my failings, or by making light of my symptoms (provided they are not too painful or serious). But deep down, I realize that I cannot hope to be spared some kinds of health problems as I grow older. Still, I continue to hope that I'll be lucky and somehow be reprieved of their onset for awhile.

When our Friends Start Falling

It is difficult enough when our friends and our peers start failing, but there is always the possibility of recovery, even from serious illness; or at least survival, even in diminished condition. So when my friends are ill, I worry and I sympathize, but I expect, and even assume, that they will recover somehow. Perhaps to expect otherwise would be to acknowledge that all of us, myself included, are at mortal risk.

It is not that death has been such a stranger in our lives but, until now, it has mostly happened to the older generation, not our own. Most of us have seen our parents die, along with other members of their generation. We were able to accept the loss, even while we mourned, because it was the natural order of things that the elderly die first; we were still the second, and "safe," generation. But now that death has begun to invade our own generation, we feel a sharpened sense of loss, as well as a threat to our personal well-being.

Although "death is the most predictable fact of life," we are almost never prepared for it, especially when it happens to someone who is important to us. If a friend dies suddenly, we experience shock, but supposedly, there is consolation in the fact that there was no prolonged suffering. But there is also no time for transition, for preparing ourselves, or for saying good-bye. Yet, even if a friend's death is gradual and expected, being prepared for it doesn't necessarily make it easy. It has been observed that "every death is shocking in its own way." The fact is that there is no way of predicting how we will react, and there is simply no "good" or "easy" way to lose someone you care about.

I live in a retirement community in which illness, hospitalization, and even death are not uncommon occurrences, especially as our population grows older. In fact, our monthly community newspaper includes a regular column entitled "In Memorium." Every issue contains its list of names, including people I know! Indeed, I have gotten to the point where I no longer send out "get well" cards promptly when I hear about an acquaintance being ill or hospitalized, just in case I may have to send a condolence card, instead! I now even read obituary columns in the local newspaper, which I never did before, and I sometimes notice, with shock, that the birth years of the deceased are the same as mine!

I find that my calendar is as filled with condolence calls as it is with as social calls. But no matter how frequent these visits, and no matter how long or how little I have known the deceased, I never feel comfortable giving condolences. If the deceased is only a casual acquaintance, I am uncomfortable, albeit courteous, with acknowledging someone's grief, or expressing any of my own. On the other hand, if it is someone I have known well, or for a long time, I am still uncomfortable, no matter how genuine the emotion I feel. This is because I do not know what to say that is helpful or meaningful because, after all, nothing I can say can alter the reality.

Even as I feel some of the survivor's grief, I also feel some sense of guilt. Among the friends and acquaintances of our generation, it is most often the husband who dies first, and the widow whom we visit in our condolence call. I sometimes wonder whether the new widow feels any resentment, even if unconsciously or involuntarily, because my husband and I have survived together; we were, after all, of similar age and history. Even if we were lifelong and devoted friends, and even if I know we did all we could, somehow, I still feel guilty when a friend dies, even if nothing blaming was said or intended. I know that the situation could have been reversed, and that I am simply fortunate in the face of another's misfortune. It is not that I especially deserve the gift of added life; I have just been lucky enough to receive it.

Gradually, over time, I have begun to learn the special vocabulary and etiquette of condolence, of how to say words that will acknowledge the loss without being too upsetting to the mourners, or to myself. I have even become accustomed to the practice of providing refreshments and socialization at condolence visits; I realize that we survivors are simply trying to defy death through life-giving activities.

When we lose a friend, we lose many things. We miss the actual presence of the person, but we also miss the relationship we enjoyed together, and the special part he or she played in our life. We know we may make other friends, but this particular person is irreplaceable, and all the situations in the future will now be different without that presence. We also lose the sense of solidarity and safety we felt, as members of the same generation facing our aging and mortality together. The death of a friend forces us to preview our own; with that loss, our own lives are diminished.

Sometimes the death of a close friend seems so unreal and so unacceptable to me that I persist in believing that he or she is only away, and will eventually return. I know that this is only my own way of trying to deny what has happened, because denial seems the only way I can protest. However, this protest and denial cannot succeed because, as the saying goes "death is not optional." I also know that, although we never become totally accustomed to loss, we ultimately learn to live, the best way we can, with whatever and whomever is left to us. The truth is that, in spite of all the pain of being a survivor, I know I would rather be one than not.

End of the Road

Most of us can still remember how it felt when we were awarded our first driver's license. If we were adolescents at the time, the license seemed to represent permission, not just to drive, but to be emancipated, to grow up. If we were already adults at the time, the license meant not just more mobility, but more convenience and independence.

Giving up driving, then, represents the loss of these freedoms. It does not usually occur all at once, or in a particular single event, unless we suffer some sudden illness or injury. Instead, it takes place gradually, over time, in a series of small, but steady erosions, with a mixture of steps and a mixture of feelings.

Usually, the first step is to stop driving at night. It's not that night driving is too strenuous for us, but it is too stressful. We must strain our vision to adjust to the glare of headlights; and our sense of direction must strain as darkness makes places look different. The next step is to stop driving on the freeways at all, even in daylight. It seems as if the pace and pressure of freeway traffic have speeded up at the same time as our reflexes and reactions have slowed down. This means that we are now limited to driving only in our own neighborhoods, and usually only for necessary tasks, because these trips are brief, familiar, and feel safe. But the less we drive, the more limited our lives become, shrinking the circle of our activities, acquaintances, and interests.

Finally, we lose, or let go of, our driver's license altogether. We may fail the renewal exam because we cannot remember the answers to the written questions, or we cannot pass the vision test, or because we are too nervous on the road test. We may even fail to take the exam at all, because we fear failing it.

Frequently, we try to minimize, or rationalize, what is happening. We try to reassure others, and ourselves, that driving really is not important anymore; or it is not worth the bother and expense of car upkeep; or we really want to give our cars to our grandchildren who are now coming of age. Although we may pretend that giving up driving is not much to give up, deep down, we know that what we are losing is more than a license!

Not driving anymore affects not only our mobility, but also our marital relationships, as older women and older men deal with this change in different ways. For most women, it is more of an inconvenience than an injury to the ego. Perhaps this is because our generation of older women was accustomed to having someone else (namely, our fathers; and then, our husbands) do the major family driving, while we women did the short-term, social, secondary driving. Now it means that we must become totally dependent (not partially or periodically, as we formerly did) on our husbands for transportation. Often, it also causes conflicts, as one or the other must forego individual interests and be forced into shared activities. Yet despite these problems, giving up driving is a transition, but not necessarily a trauma, for women of my generation, because we have been accustomed to having our driving limited, anyway.

For older men, on the other hand, giving up driving seems more drastic. Perhaps this is because so much of their sense of self is associated with being strong and self-reliant. Perhaps this is also why men are usually more resistant to giving up driving until finally compelled by incontrovertible physical or health failures. This means that their wives, who were usually the secondary or less-experienced drivers, now become the family chauffeurs, responsible for all medical, social, business, housekeeping, and personal transportation. For many of these women, it was their husbands who once taught them how to drive; now, their roles are reversed.

Men often respond to this changed circumstance in one of two opposite ways: Some have found it so difficult to be dependent that they avoid asking for transportation favors, and therefore, become increasingly, and resentfully, housebound. Others demand frequent chauffeuring by their wives, but become hypercritical about their wives' skill behind the wheel, as if this provides them with some substitute sense of power.

Although reactions may differ, the realities are similar for all of us, men and women alike. Giving up driving not only limits where we can go and how we can get there, but also what we can do and what our lives can be. Being able to drive provided us with the convenience of mobility, as well as with the assurance of independence. At the beginning, we experienced a surge of power and freedom; now, we suffer the reverse, because we know that we are also losing part of our independent selves.

Yet we are all aware of those older people with failing vision, reflexes, and judgment, who still insist on driving, even though they should not, and who constitute a menace to themselves and others. Therefore, we hope that we will be able to recognize the new reality of the road for ourselves, when the time comes; and be able, even though reluctantly and regretfully, to let go. We hope that we will know when it is right for us to move out of the driver's seat and into the passenger's seat for the remainder of our travels.

My Place or Your Place?

When we were children, independence was our dream; when we were adults, independence was our reality; but now that we are older, loss of independence has become our fear. Over the years, the places we live in, as well as the ways we live in them, reflect these changes.

We start out as children, living in the home of our parents, upon whom we are dependent and who are in charge of our lives. Then, in our young and middle years, we live in our own homes, houses, or apartments. But whatever the physical structure, we are now completely independent and in charge of our homes, our families, and our lives. Finally, in our later years, we find that little by little, out of both convenience and necessity, we begin to relinquish some of our responsibilities and independence to others. It is not that we revert to becoming dependent, as we were at the beginning, but our homes, as well as our lives, go through a process of transition and attrition.

The early moves, from childhood home to adult home, were ascending steps into maturity. The first descending move into age occurs when we exchange our home in the outside community for one in a retirement community. However, though we know that this is the first move related to our aging process, it still feels like a positive change, because it seems to offer all the advantages, and none of the disadvantages, of our usual lifestyle — independence and privacy, without maintenance or responsibility.

But along with giving up work and worry, we also give up some control. We delegate external management to an outside organization in order to enjoy freedom from care and physical labor. We allow the organization to make rules and regulations governing permissible or impermissible behaviors in order to enjoy comfort and convenience. We give up diversity of population in order to enjoy peer companionship; and we give up being part of the larger, unfenced world in order to enjoy security. Yet, overall, we usually do not feel such a great difference between living in a private home and living in our home in a retirement community.

However, the cloud that hangs over us is the shadow of the next step, because we know that any further move we make will take us irreversibly further from independence. The retirement community in which I live is now 15 years old, and the original residents, who moved there in their 60s or 70s, are now in their 80s, or even early 90s! At this stage, some of them are finding independent housekeeping too difficult due to illness, injury, or merely the debility of age; and they are moving, or thinking about moving, into so-called "retirement" homes. It is this move that is the big leap into old age because, not only is it based on old age, as is the retirement community, but it is based on *infirm* old age. Moving into a retirement *community* means giving up things that we no longer want to be bothered doing for ourselves; but moving into a retirement *home* means giving up things that we are no longer *able* to do for ourselves.

There are many different kinds of these senior residences, ranging from small, converted private homes to medium-sized board-and-care facilities to large hotel-type arrangements. They also range from Spartan simplicity to lavish luxury, and are usually called by various euphemistic names, such as senior "villa" or "hotel" or "gardens." But no matter the name, size, or style, it is no longer your own home, but one you share with other aging strangers who also can no longer manage by themselves.

You share a communal living room, a communal dining room, a communal porch, and a communal foyer. The space that is privately yours shrinks from a house to a tiny suite, or even one room. You eat food prepared and selected by others at meal times designated by others; you participate in activities planned and managed by others; you follow schedules defined by others.

These retirement homes have recently begun to encourage new residents to bring their own furnishings and decorations with them, to make them feel more "at home." This can be helpful, but also painful. First of all, since the living space is now so much less, many belongings must be left behind or disposed of. Even so, there is a tendency to crowd too much into too little room, not because all these things are really needed, but because the more you hold on to what is yours, the more independent you feel. However, the comfort of what is familiar may also be a reminder of what is different, and the possessions that remain may also recall those that are now gone. The truth of the matter is that, deep down, you know that no matter how much you try to feel "at home," you are really *not* home; you do not *have* your own home anymore!

I remember years ago, when my parents moved into a retirement home, I felt a sense of loss for myself, as well as for them. They were losing their independent home, but I was losing my sense of independent parents who were able to offer me their home as a place to come to. It used to be that my experience with retirement homes was connected with my parents' generation, as I visited them or their friends who lived there. But now I am seeing my own neighbors and acquaintances moving into these homes, or considering doing so, and I am visiting my own peers who live there. How can it be that it may be my turn next?

Just as retiring from work seems to represent the onset of aging, giving up one's independent home represents the losses that aging brings. I have already passed the first milestone, but still feel a long way from the second. I am comforted to still see that, in my retirement community, most of the people continue to manage well — or, at least, well enough. Of course, I have no way of foretelling my future, but I certainly hope that, in spite of my complaints and constraints, I will be able to be in charge of my own kitchen, my own home, and my own life, for a long time to come.

Part VII

Learning Our Lessons

You Can't Go Home Again, or Can You?

"Home" can mean many different things, from an actual physical place, or a feeling of recognition and familiarity, to a mixture of memories of old sights, sounds, and scenes. The place where we live now is our house, but our *home* is where we came from.

At this stage of our lives, we begin to engage in what psychologists call "the life review process." This means journeying into our past in order to understand our present. It means re-examining the steps, or mis-steps, that led us to where we are now, as well as what we are now. As we deal with our own present aging, we also have a need to try to clarify our past.

In her recent autobiography, author Gloria Steinem advised that there is no single way of re-entering the past. Each of us embarks on our own private journey by following our personal clues and paths backward into our past lives.

This process involves risks, as well as the ability to remember. Unexpected reminders can tear down walls built up around certain parts of our lives. Our bodies even seem to retain special memories of their own, triggered by tastes, sounds, and smells, without conscious mental control. We may remember selectively, filtering things through the way we prefer things to have been. Sometimes, we insist on certain memories, even though we know otherwise; or we feel obstinately positive about some, and vague about others; or, in recollection, we see things we were not aware of, or impressed by, at the time. Or we may look back at crossroads in our lives and wonder what would have happened if we had pursued different paths. I wonder, sometimes, how we can be so forgetful, and yet so nostalgic, at the same time. The past seems both half-remembered and half-imagined, and so it becomes both real and unreal.

Sometimes, going "home" can be a literal, as well as a figurative, journey — an actual visit to places in our past. This can carry different kinds of risks and rewards. First of all, these physical places may no longer exist. Secondly, if they do, they will have been transformed by time. Our childhood homes stood so large in our eyes because we were so little; and every place we saw seemed new and imposing, because we were so young and impressionable. We may have pictures in our minds of the house we lived in, the school we attended, or the streets we traveled, but it is doubtful whether these places could be located from our remembered descriptions.

It is not necessarily that time has played tricks with our memories. The physical structures themselves may have been altered over the years. They may have been enlarged, rebuilt, or refinished. And the surrounding scene, as well as the structure, may have changed. It may be more built-up, or more run- down, thereby changing the entire setting. And even if these physical changes have not occurred, we now view them through a different perspective than we did when we were younger. As a result, what we find may not be what we remember, and what we remember may not be what we find.

Recently, my husband and I made such an actual journey back to the place of his childhood, more than half a century, an ocean, and a continent away! The small Eastern European town in which he grew up had been destroyed by war, and rebuilt in the peace that followed. New houses had been built, and even the street numbers were changed, but the house he had lived in was still there, and still recognizable.

The present occupant invited us to come inside, to wander through the rooms, to take pictures. My husband began excitedly trying to describe to me how the rooms had looked, how the furniture had been arranged, how there had been a garden where the garage now stood, and flower pots instead of television aerials on the balcony. After awhile, he grew more quiet and less certain of how things had been so long ago. Finally, we thanked the host, walked around outside for a final view, and then left.

Afterward, my husband reflected, with both puzzlement and relief, that he knew it was the same building he remembered, but it did not seem like the same home to him. Yet I do not think the visit was a loss or a waste, even if what was found was not as my husband remembered, or imagined, it to be. The important thing is not to try to reclaim the past, which is not possible anyway, but to confirm that it existed. After all, the past should be a memory, not a memorial!

Novelist Thomas Wolfe titled his most well-known book *You Can't Go Home Again*. But it seems to me that this warning is only true in a partial, or literal, sense. Certainly it is true that we cannot journey back to the children we once were; or to a particular place or time which are now gone or changed; or to the way we used to be in that place or at that time. But in another sense, part of our childhood "home" has become part of us. We carry within us, knowingly or unknowingly, voluntarily or involuntarily, the most meaningful facts and forces from our earlier lives. In other words, even though the years may have taken us out of our old home, they have never completely taken our old home out of us.

Review and Remembrance

Older people are often alternately accused of not remembering enough and of remembering too much. These criticisms are true, and yet not true. Yes, we sometimes miss, misplace, or mistake facts of the present; and yes, we find frequent reminders, references, or resemblances to facts of the past. But just because we do not dismiss the past does not mean that we deny or dislike the present.

When we were younger, we had not accumulated enough experiences, or were simply too busy living our present lives, to have the time, or the need, to remember much. Now that we are older, and have more time behind us than ahead of us, we remember a great deal, because we have a great deal to remember! Getting older does not mean that we stop thinking about our lives; in fact, this may be the time when we really start to do so.

Indeed, today, psychiatrists and gerontologists are telling us that reminiscence in older people is actually healthy and therapeutic. We could have told them that ourselves, without the professional jargon. Our past is something that is uniquely and permanently ours; and remembering it is something we can do effortlessly, and which becomes even easier and more enjoyable — not less — as we grow older.

What is it we enjoy remembering? We recall the important people in our lives who are no longer with us, or who are no longer nearby. This usually means our parents, but also includes other close relatives, close friends, mates, and classmates. We remember them as they used to be, when they played important roles in our lives. We remember ourselves as we once were — younger, thinner, stronger, straighter. We also remember milestone events in our lives, not necessarily earthshaking or major achievements, but experiences that especially touched or taught us. It is not that we want to revive or relive them; we just don't want to lose them!

It's possible that what we remember about these people and events may be selective, or filtered by time, leaving mostly pleasant pictures. But it seems to me that it's not the absolute accuracy that matters, but the *meanings* these memories hold.

How do we remember? We remember by ourselves, or with others, and in different ways. We remember frequently with our peers, with whom we can communicate so easily. It seems that we have shared so many similar experiences in our lifetimes that we speak a special short-hand language with each other. When one of us recounts an experience, there is immediate understanding and recognition.

We also remember with our families, most often with our grandchildren. Somehow, they seem to have greater curiosity and patience than our own adult children for learning about what used to be. This is particularly true for hearing stories about their own parents as children and, even more unbelievably, about ourselves as children. We enjoy offering them our carefully saved souvenirs, and we hope that these pictures of our past will be preserved by them for the future.

We remember by ourselves whenever we keep our diaries, write our histories, trace our family trees, or record our recollections. Sometimes, we just remember in solitary, but not necessarily sad, reflection.

Why do we remember? We remember for many different reasons, and for our own sakes, as well as for others'. First of all, the speed of time seems to accelerate so much when

225

we get older that we wonder where the years went. So we remember just to try to preserve the time that is passing by.

Secondly, we remember in order to validate what we once did and once were. We know that we are no longer those same persons, and never will be again; and there are things about ourselves in the past that no one present has ever seen or suspected. Now we want to share them with others, or simply confirm them for ourselves.

We also remember to remind ourselves of the paths we have traveled from the past to the present, to try to better understand what we could not understand earlier in our lives. We want to finally bridge the gap between what we wanted when we were young and what we have now. To sum up, we remember for the sheer pleasure it brings us, as well as for proof to others that we have lived, before we no longer do!

Certainly there can be unhealthy or excessive remembering. We can dote on the past to the disadvantage of the present. We can, consciously or unconsciously, indulge in wishful thinking or idealizing. We can long for what is gone, and try to relive it, or repeat it, or remain there. Yes, it's true that some people do get stuck in the past.

But for most of us, remembering is neither senility nor sentimentality, and certainly not a substitute for the present. It is a healthy and reasonable exercise, because we know that the past is the parent of the present, and we recognize and respect the difference. I disagree with those who despair that older people "live more by memory than hope." Except perhaps for a small percentage of the very old or the very ill, I think this statement is both unkind and untrue. Most of us, when we grow older, are able to enjoy both memory and hope. Indeed, as one older person explained, "I do not live in the past; it is the past that lives in me."

Letting Dreams Go

I recall reading that, when we are young, we dream young dreams; and when we are old, we dream old dreams. That says two things to me: 1) there are dreams that are appropriate for both youth and age, and they are different; and 2) no matter how old we are, we still have our dreams!

I remember that when I was young, my dreams had to do with achievement, success, and power (what someone once called "trumpets and triumphs"). At first, I dreamed only of my own personal accomplishments; when I married, I dreamed about what my husband would accomplish; then, when I had children, I dreamed about the kinds of life successes I wanted for them. Now I know that none of these will come true. I will never write "the great American novel;" my husband will never be president of the company; nor will my daughters ever be close in location or lifestyle.

Today, I have different kinds of dreams; no longer of power, but of ordinary pleasures; no longer a quest for achievement, but for quiet contentment. It's not so much settling for less, but moving on to something different. Psychologist Roger Gould observed that children mark the passage of time in their lives by changing their bodies, adults by changing their minds!

Of course, there is some feeling of loss about things that are finished, or are no longer possible. I will never again be a productive, full-time working person; or again be a child to my parents; or be a new parent to my children, or a new spouse to my husband (in the same bright, beginning way).

Those roles and tasks are finished, for better or for worse. I cannot redo them; I can only remember them.

I know now that the gap between dream and reality will never be completely closed, and I will never have the opportunity to do certain things, or to undo certain mistakes in my life. I have to accept what I am, not what I once wanted to, or might have, become; and this is also true of my expectations of others — of my husband, and of my children. If I cannot let go of what cannot be anymore, then it ceases to be a dream, and becomes a frustrated fantasy, instead.

In a recent poem about growing older, Judith Viorst wrote:

"Getting old is tough, but we find a few compensations ... and happiness arrives in new disguises."

This means that letting go of the dreams I had for myself and for others may not be a loss, but rather, can be freeing and refreshing. This has been called "de-illusionment," which is not the same as "dis-illusionment." In this process, we give up earlier illusions without pain or pity, and we proceed to new, "adult dreams" that are still achievable. We may redefine success, or settle for partial accomplishments. We may turn our attention to other aspects of ourselves that have been ignored or silenced till now. We may see options or viewpoints that were not visible to us before. We may have fewer possibilities, but we have more perspective. In other words, growing older can also mean growing up!

I remember too that, in the past, there was pleasure for me in having those dreams, and there was pride in whatever I was able to accomplish. Now, together with regret and remembrance, there is also some relief. Letting go of what never can or will be leaves us with what we are! This doesn't necessarily mean settling for a spoonful instead of a plateful, but more like changing the menu!

It is now time for our adult children, and soon our growing grandchildren, to pursue their dreams. We do not bequeath our ambitions to the new generations, but they will probably have similar goals of accomplishment and acclaim. We wish for them that they may accomplish more than we did, and most of what they want! We envy them a little for still having the time for, and belief in, the possibility; but we also appreciate our new luxury of emancipation and are willing to enjoy vicarious victories.

It seems to me that giving up old hopes and dreams must be most difficult for those who have the most regrets about their lives. We all have some regrets about "roads not taken," plans not fulfilled, or temptations resisted. I don't think any of us, in honesty, can claim that it all turned out the way we really wanted, or expected, it would. But we need to finally accept what's past without being preoccupied with it, and to banish blame for what was not able to be. It's time now to remember the past without regret, to think of the future without fantasy, and to allow ourselves the possible pleasures of the present.

Wills and Won'ts, or
"What Legacy are We Really Leaving?"

A legacy is defined in the dictionary as a gift that one generation bequeaths to the next. An estate is the sum total of material possessions, but a legacy contains intangible bequests, as well. It is our hope to be able to leave both to our children.

As we grow older, we begin to think more and more about what will remain after us, or be remembered of us. It is not so much something we worry about, but something we wonder about. We think about this especially as we watch the generations in our family begin to unfold, as our adult children approach middle-age, and our grandchildren approach young adulthood. We realize, with a certain amount of shock and disbelief, that we are now the older generation and, in time, will become the absent generation.

With this in mind, my husband and I have recently begun to take serious steps, both practical and personal, to define both our estate and our legacy. In other words, we have begun to "put our house in order." We have updated our wills; we have tried, with legal advice, to establish the most financially advantageous arrangements; we have made an accurate inventory of our possessions; we have left clear and careful instructions; we have tried, in every way we could think of, to pre-plan, so that there would be the least amount of burden to our children.

There is nothing morbid or melodramatic for us in doing any of this. On the contrary, it gives us the feeling of still being in control of our lives. It also offers us an opportunity for a final act of parenting, by protecting our children from these troubling tasks. In addition, we feel a certain satisfaction, if not pleasure, in being able to tie up loose ends and bring things to closure.

However, when we try to discuss this with our children, they are unwilling, uncomfortable, or unresponsive. They usually attempt to ward off the subject through distractions or denials. I know that it is not because they don't care about us or don't worry about our mortality; in fact, it is precisely because they *do* that this subject is so difficult for them. On the other hand, my husband and I feel pleased and a little proud that we will be able to leave an estate to our children that can materially contribute to their future well-being.

We have also tried to be scrupulously equitable with our estate, without any special regard for individual need or perceived merit. We have several reasons for this: First of all, we feel that it is now long past the time to offer rewards or punishments, or try to balance emotional accounts. Also, it seems both futile and wrong, at such a time, to render unilateral judgments without opportunity for response. Finally, we know that the possessions we leave to our children may become more than the objects themselves; that they may take on a symbolic value and become reflections of our feelings. I have seen many situations in which survivors were anguished by how little they were left, or puzzled by how much. So we want our final statement to our daughters to tell them that we loved them both, perhaps not always wisely, but always well, and always equally.

We also want to leave them a personal legacy, or an "ethical" will; this may be the last chance we have as parents to teach our children how to live. In other words, we want to bequeath not just objects of value, but also a *sense* of values. Hopefully, we've been doing this all along; but if our children are willing to accept our property, they also ought to be willing to understand our principles. I suppose that, deep down, we hope that they will not only *remember* our beliefs, but also *share* them. It seems ironic that, now that we have finally begun to learn some of the important answers, we are not certain that we will be asked the questions!

Our concern about leaving a moral legacy is particularly powerful in regard to our grandchildren. With our children, we may be too close or too late; but it is our grandchildren, whom we see as the real future, that we especially want to know about us. In my faith and in my family, children are named after close relatives who are no longer living as a reminder that each of us is a link in the chain of generations. My own grandchildren are named after my mother and father. It's strange to imagine that someday, there may be a child in my family who will carry my name!

I suppose that the bottom line of our concern is not so much being understood or appreciated, but being remembered. I hope our children will remember what my husband and I were really like, with some sympathy, but without sentimentalizing or sanctifying. I hope they will remember my good cooking and my bad puns; and that my husband was often hard-headed, but always soft-hearted. I hope they will remember that he was the one who added excitement and intensity to our lives, while I was the one who was the steady center through family storms; so that, together, we made a complete couple. I hope they will remember us often enough, and mostly with smiles!

Epilogue — Summing Up

The summing up process is a part of life, just as it is a part of mathematics; at certain points, we try to "add things up," so we can evaluate the results so far. Summing up does not mean coming to a finish, or to any final answers, but only to a particular point of review and reflection. In fact, we cannot really proceed much further until we first understand how we have come this far.

For me, this year in my life is just such a watershed point, as I come to the end of this book and to the beginning of my biblically allotted span of years — "three score and ten!" It's not that I don't look forward to more years ahead of me, but I also need to look back, at the years behind me.

This process of summing up is sometimes called "life review;" but whatever name and whatever time frame we use, it is really a final attempt to learn who we are. In doing this, we do not attempt to recapture our lost youth, but to recognize what we have learned from it. When we were young, we didn't worry very much about our age; now that we are old, we worry about it a great deal! As the saying goes, "growing old is a life's work," so I would like to do the best job of it I can.

I have learned a lot of things along the way to becoming older, and not all of them were lessons I anticipated. When I was younger, I thought there were solutions to all problems, and answers to all questions; and I expected to find them out, and to find them all. Now I see that there are usually many sides to every circumstance, and that some questions have no certain answers, or even any answers at all.

I have learned that, although there are no absolutes, there are limits to what we can do, what we can know, and what we can be. When we were young, we didn't even know what we couldn't do; it was not until later in life, after enough things that were possible actually happened, that we could realize what was not possible. As a result, I have learned that there is no way we can end our lives with all problems resolved, all needs satisfied, and all loose ends neatly tied together.

There were things I wanted to do, or have, or be; I expected that I would deserve them, and I hoped that I would achieve them. Now I find that some of them, I really didn't want after all; and some of them, I will never be able to have, no matter how much I want them. So I have had to redefine my expectations — sometimes with regret, and sometimes with relief — so I could live with the way things did not turn out, as well as with the way they did. Judith Viorst summed it up in one of her verses:

"It's not what I called happiness
When I was twenty-one,
But it's turning out to be
What happiness is."

I have found out that other people are no more predictable, powerful, or perfect than I am; and that is both frustrating and consoling to me. But it also means that we cannot hold one another endlessly responsible for only being what we are, and not what we want each other to be. Comedian Charles Grodin, in a serious moment, observed, "We are all only imperfect people in an imperfect world."

Not all of these lessons have been easy to learn, but they have had to be accepted. As TV commentator Linda Ellerbe warned, "The only thing harder than learning from experience is *not* learning from experience!"

Although many things in my life have turned out differently than I expected — some for the better, and some not — enough has been good or, at least, good enough. I am grateful for whatever I was able to do, even though I am regretful about whatever I was not able to do, especially now that I know there will not be time. I have learned a great deal, much of it in this last stage of life. I wish I had time to learn more, and I wish I had learned some of these things earlier, so that I could have had more time to make use of them. It's too bad that it took so long for me to get to this point, but the truth is that I probably could not have gotten here any sooner.

I sometimes wonder what I would do if I had it to do all over again. If I could go back, what would I change? Certainly, there are some things that I wish had been different (given what I know now that I did not know then), but I am not sure what else I would have done at the time. Hindsight may be interesting, but it is not necessarily helpful.

It seems strange to realize that I am now the way I remember my elderly parents to have been, and my children are now the way I remember myself to have been. It is even stranger to think of life going on without my being there to be part of it, as I know must someday happen. It saddens me to think of my grandchildren growing up to lives I will not see, like planting seeds that I will never be able to enjoy seeing come to flower. Yet, at the same time, knowing that there will be new generations to follow is one of the greatest comforts of all.

Judith Viorst, in another poem she wrote to her family, said to them:

"Before I go,
I'd like to know that I've been more of a joy
than a pain."

That is exactly the way I feel, as I think about my own family. I know that I have not always been easy to live with, that I have sometimes tried too hard, or even cared too much. It was not that I wanted so much *from* those I loved, but that I wanted so much *for* them. I would like the people I love to know that I tried to do the best I could with whatever I knew at the time. I would also like them to know that I am glad we have been able to put up with each other all these years. I am glad they have been part of my life, and I hope that they are also glad that I have been part of theirs.

I know that there are some who complain that they are old too soon, and for too long; and that there are others who complain that they are not old long enough! I suppose it depends on how much one misses what came before growing old, and how much one fears what may come after. As for me, I know I would rather grow old than not, and I am willing to accept the changes that aging must bring, even including any "necessary losses." After all, the only alternative to experiencing loss in one's life is not to have anything, or anyone, to lose!

So, if I had to sum it up so far, I would say that growing older is a time when things begin to fall apart in ways they never used to, and things also begin to come together in ways they never used to. In other words, age may not be a laughing matter, but it certainly is not a crying matter, either!

INDEX

Other Titles from Elder Books

Reminiscence: Uncovering A Lifetime of Memories
by Carmel Sheridan
Older adults get in touch with things that were important to them through reminiscence. The process of reviewing memories helps validate who they are and builds their sense of identity. At a time in their lives when older adults may feel most vulnerable, isolated or lonely, reminiscing and memory-sharing helps restore confidence and self-esteem and acknowledges their contribution to life. In this way, reminiscence can improve mental, emotional and sometimes physical well-being. Older adults have a powerful resource in reminiscence and this book shows how you can help them utilize this resource.   **$12.95**

Surviving Alzheimer's: A Guide for Families
by Florian Raymond
Easily digestible, this book is a treasure house of practical tips, ideas and survival strategies for the busy caregiver. It describes how to renew and restore yourself during the ups and downs of caregiving, and shows you how to take care of yourself as well as your family member.   **$10.95**

In Sickness & In Health: Caring for A Loved One with Alzheimer's by Dr. William Grubbs
In Sickness & In Health is a caregiver's courageous tale of caring for his wife who had Alzheimer's disease. Dr Grubbs talks candidly about the agonizing decisions he was forced to make during his wife's slow decline. He addresses the problems of caring for a spouse at home, and deals frankly with the dilemma of placing a loved one in a nursing home. Drawing on first-hand experience, he describes how caregivers can keep themselves mentally and physically healthy, and reach out for much-needed support. Families, support groups and medical personnel will benefit greatly from reading this touching and insightful chronicle.   **$12.95**

ORDER FORM

Send to:

Elder Books PO Box 490 Forest Knolls CA 94933
Ph. 1 800 909 COPE (2673) Fax. 415 488 4720

QTY	TITLE	PRICE/COPY	TOTALS
____	Finishing Touches	$12.95	____ . ____
____	Reminiscence	$12.95	____ . ____
____	Surviving Alzheimer's: A Guide for Families	$10.95	____ . ____
____	In Sickness & In Health	$12.95	____ . ____

Total for books . . . . _____ . ____

Total for shipping . . . . . . _____ . ____

Total sales tax . . . . _____ . ____

Amount enclosed . . . . . . _____ . ____

Shipping $2.50 for first book, $1.25 for each additional book; California residents add 8.25% sales tax.

Name _____

Address _____

City _____ State _____ Zip _____